CLARENDON LIBRARY OF LOGIC AND PHILOSOPHY

General Editor. L. Jonathan Cohen, The Queen's College, Oxford

THE METAPHYSICS OF MODALITY

Also published in this series

Quality and Concept by George Bealer
The Probable and the Provable by L. Jonathan Cohen
The Diversity of Moral Thinking by Neil Cooper
Interests and Rights: The Case Against Animals by R. G. Frey
The Logic of Aspect: An Axiomatic Approach by Antony Galton
Relative Identity by Nicholas Griffin
Ontological Economy by Dale Gottlieb
Equality, Liberty, and Perfectionism by Vinit Haksar
Experiences: An Inquiry into some Ambiguities by J. M. Hinton
Knowledge by Keith Lehrer
Metaphysics and the Mind–Body Problem by Michael E. Levin
The Cement of the Universe: A Study of Causation by J. L. Mackie
Truth, Probability, and Paradox by J. L. Mackie
The Nature of Necessity by Alvin Plantinga
Divine Commands and Moral Requirements by P. L. Quinn
Simplicity by Elliott Sober
The Logic of Natural Language by Fred Sommers
The Coherence of Theism by Richard Swinburne
The Emergence of Norms by Edna Ullmann-Margalit
Ignorance: A Case for Scepticism by Peter Unger
The Scientific Image by Bas C. van Fraassen
What is Existence? by C. J. F. Williams
Works and Worlds of Art by Nicholas Wolterstorff

THE METAPHYSICS OF MODALITY

GRAEME FORBES

CLARENDON PRESS · OXFORD

Oxford University Press, Walton Street, Oxford OX2 6DP
London New York Toronto
Delhi Bombay Calcutta Madras Karachi
Petaling Jaya Singapore Hong Kong Tokyo
Nairobi Dar es Salaam Cape Town
Melbourne Auckland
and associated companies in
Beirut Berlin Ibadan Mexico City Nicosia

Oxford is a trade mark of Oxford University Press

Published in the United States
by Oxford University Press, New York

First published 1985
Reprinted (new as paperback) 1986

British Library Cataloguing in Publication Data
Forbes, Graeme
The metaphysics of modality.—
(Clarendon library of logic and philosophy)
1. Modality (Logic)
I. Title
160 BC199.M6
ISBN 0-19-824432-0
ISBN 0-19-824433-9 Pbk.

Library of Congress Cataloging in Publication Data
Forbes, Graeme.
The metaphysics of modality.
(Clarendon library of logic and philosophy)
Bibliography: p.
Includes index.
1. Modality (Logic)—Addresses, essays, lectures.
2. Metaphysics—Addresses, essays, lectures.
I. Title. II. Series.
BC199.M6F65 1985 111'.1 84-16674
ISBN 0-19-824432-0
ISBN 0-19-824433-9 Pbk.

Typeset by Joshua Associates, Oxford
Printed in Great Britain by
The Alden Press, Oxford

Preface

I HAVE written this book with two purposes in mind. First, I wanted to produce something which a reader could use as a means of entry into the area of analytic metaphysics concerned with modality. Secondly, I wanted to make a contribution to the literature in this area which would be of interest to those working in it. A reader of the latter sort may recognize that much of the content of this book has appeared in rather disconnected fashion in various journal papers over the last few years, but it is my hope that even those familiar with those papers will find something new here in what I say about topics covered in them, and something worthwhile in what I say about topics on which I have not previously written. I also hope that the theory of individual essences with which this book is mainly concerned benefits from being presented as a whole in a single place.

I have tried to make as few demands for prerequisites as possible on the reader who would like a way into the general area of the metaphysics of modality, so the only real requirement for reading this book is familiarity with modern logic at the level at which most elementary symbolic logic courses are conducted. Specifically, I have assumed that the reader understands the distinction between valid and invalid arguments (argument-forms, sequents) in propositional and predicate calculus, and knows something of how this distinction can be characterized. But I have not assumed any previous acquaintance with modal logic, and the first two chapters of the book are therefore given over to introducing the reader to this subject. I have done this on what secretive bureaucrats call a 'need to know' basis, and so I have made no attempt at completeness or even a very high degree of rigour in my presentation (specifically, I have included nothing about axiomatic formulations of modal systems, since the topic is irrelevant to my philosophical purposes). Sometimes, I have tried to repair the effects of this casualness in the longer footnotes, most of which have been written for enthusiasts, and are therefore less accessible than the main text; but this is the way of the world with footnotes.

I should also say something about my use of English quotation marks. When these surround a complex expression, they may either be performing their usual function of forming a name of that expression, or they may be functioning as quasi-quotes, depending on whether, in the context, it is more natural to think of the complex expression as a formula of some imagined formal language, or as a schema for such formulae; in the Appendix, corner quotes are used for quasi-quotation, but they would have given a cluttered look to the main text. Concomitantly, atomic sentence letters 'P', 'Q', etc., are quoted or unquoted depending on whether it is more natural to think of them as belonging to the lexical primitives of some formal language or as names of particular, say English, sentences.

Many people saw parts of this book in typescript form and gave me helpful comments, but I will not give a long list of names here, although I would like to thank an anonymous reader for the Press. However, a number of people played perhaps a more crucial role through the influence they had on me at a time when my ideas were being formed, and in this connection I ought to mention Martin Davies, Michael Dummett, Kit Fine, David Kaplan and David Wiggins. Most especially, I must acknowledge my long-standing and continuing debt to Christopher Peacocke, who was successively my dissertation supervisor and colleague; I do not much like to dwell upon the thought of what this work would have been like without his influence, encouragement, and help.

New Orleans G. F.
October 1983

Acknowledgements

CERTAIN passages of this book have been drawn from my previously published papers. For their co-operation in this respect, I wish to thank the publishers of the following journals: *Philosophical Studies*, for material from 'Origin and Identity', 37 (1980) 353–62, copyright © 1980 by D. Reidel Publishing Company, Dordrecht, Holland; *The Journal of Philosophical Logic*, for material from 'On The Philosophical Basis of Essentialist Theories', 10 (1981) 73–99, copyright © 1981 by D. Reidel Publishing Company, Dordrecht, Holland, and 'Physicalism, Instrumentalism and the Semantics of Modal Logic', 12 (1983), copyright © 1983 by D. Reidel Publishing Company, Dordrecht, Holland; *Synthese*, for material from 'Thisness and Vagueness', 54 (1983) 235–59, copyright © 1983 by D. Reidel Publishing Company, Dordrecht, Holland; and *Mind*, for material from 'Wiggins on Sets and Essence', 92 (1983) 114–19.

The later stages of the production of the first complete draft of this book were assisted by a grant-in-aid awarded to me by the American Council of Learned Societies, for which I would like to express my sincere appreciation.

Acknowledgements

Contents

1 PROPOSITIONAL MODAL LOGIC 1
1. The sentential operators '□' and '◇' 1
2. Invalid arguments; semantics for S5 3
3. Other systems 11
4. Incomplete circumstances: possibility semantics 18

2 FIRST ORDER MODAL LOGIC 23
1. Operators, quantifiers and invalid inferences 23
2. Semantics for quantified S5 28
3. First order tense logic 38
4. Possibility semantics for quantified S5 43

3 THE *DE RE*/*DE DICTO* DISTINCTION AND THE PROBLEM OF TRANSWORLD IDENTITY 48
1. Two kinds of formula 48
2. Quine's view 50
3. Eliminating *de re* modality 54
4. Counterpart theory 57
5. Objections to counterpart theory 64

4 METAPHYSICS FOR THE SEMANTICS 70
1. Semantics and explanation 70
2. Realism about worlds 76
3. Two problems for anti-realism 80
4. Validity: other approaches 81
5. The meanings of possible worlds sentences 89

5 A MODAL THEORY: THE ESSENCES OF SETS 96
1. Essential properties and essences 96
2. The essences of sets 100
3. The system MST 114
4. Justifying Membership Rigidity: two unsuccessful attempts 123
5. The grounding of identities and non-identities 126

6. THE NECESSITY OF ORIGIN 132
1. Kripke's thesis 132
2. An unsuccessful defence of (K) 134
3. The case of the moveable oak tree 138
4. Intrinsic and extrinsic grounding 140
5. Essences and bare particulars 145
6. The branching conception of possible worlds 148
7. A problem about identity through time 152

7. FUZZY ESSENCES AND DEGREES OF POSSIBILITY 160
1. Two paradoxes 160
2. Sorites paradoxes 164
3. The semantics of vagueness 169
4. Closgs and counterparts 175
5. Counterpart theory with degrees of possibility 180
6. Consequences 184

8. SUBSTANCES, PROPERTIES, AND EVENTS 191
1. Substances as things 191
2. Identical substances are necessarily identical 195
3. Crossworld equivalence relations 197
4. Properties 202
5. Events 205
6. Lombard's essentialism 208

9 THE JUSTIFICATION OF MODAL CONCEPTS 216
1. Non-cognitivism 216
2. Quasi-psychologism 220
3. The theory of content 224
4. The source of necessity 230

APPENDIX: Translation Schemes 238
1. Sentential Modal Logic 238
2. Quantified S5 239
3. Counterpart-theoretic S5 240
4. Reverse translation 242
5. Possibilist quantifiers 243

BIBLIOGRAPHY 248

INDEX 255

1

Propositional Modal Logic

1. *The sentential operators '□' and '◇'*

MOST people believe that in many respects things could be different from the way they actually are. We often have 'if only' thoughts—if only Jones had taken his broker's advice, he would be a millionaire today—thoughts which would lose some poignancy if the way things are, including Jones's impecunious state, is the only way it is possible for them to be. Consider, then, the assertion

(1) Jones could be a millionaire today.

A way of saying the same thing is

(2) It could be that Jones is a millionaire today

which, although more cumbersome, suggests that we might regard sentences with verbs modified by 'could' or 'could have', such as (1), as contractions of sentences with verbs in the indicative mood, these verbs occurring in a subsentence itself governed by a *modal operator*, which in (2) is the phrase 'it could be that'. We symbolize this operator as '◇' and often read it as 'possibly' or 'it is possible that'.

However, there is ambiguity here of which one should beware. On one perfectly natural way of hearing

(3) It is possible that Jones is a millionaire today

an utterer of (3) is saying that *nothing he knows* is *inconsistent with* Jones's being a millionaire today. This is the *epistemic* sense of 'it is possible that', in which it means something like 'for all that is known'. In this sense, the typical reader of this book cannot truly say that it is possible that he is a millionaire today, since he knows very well that today, like other days, he has an overdraft. But we mention the epistemic sense of 'possibly' only to distinguish it from the sense of 'possibly' with which we shall be concerned; this second sense involves the kind of possibility relevant to 'if only' thoughts, and is sometimes called the *broadly logical* sense of 'possibly'.[1] As a rough elucidatory guide, 'it is possible that P' in the broadly logical sense means that there are ways things might have gone, no matter how improbable they may be, as a result of which it would have come about that P. So in this sense it *is* true, for the typical reader, that it is possible that he is a millionaire today, just as Jones would have been if he had taken his broker's advice.

Not everyone agrees that things could be different from the way they actually are. A fatalist, for example, holds that the way things actually are is the only way it is possible for them to be; so a kind of necessity is imputed to things being that way. In fact, fatalism is usually held with respect to the future: if it is going to rain tomorrow, then it *must* come to pass that it rains tomorrow.[2] We shall return briefly to fatalism in the next section of this chapter. For the moment, let us just note how we can use either the notion of possibility or of necessity to express the fatalist doctrine. We can say that how things are is how they *must* be, or that it is *not possible* for them *not* to be that way. Thus, a necessary state of affairs is one whose failing to obtain is not possible, so we can define necessity in terms of possibility:

(4) 'It is necessary that P' means that it is not possible that not-P

or in symbols, using '$\Box$' for 'it is necessary that',

(5) $\Box A =_{df} -\Diamond - A$.

As the reader may suspect, we can do the same thing the other way round, since a state of affairs is possible if and only if (henceforth 'iff') it is not necessary that it does not obtain. Again in symbols,

(6) $\Diamond A =_{df} -\Box - A$.

In these definitions, we are introducing '$\Box$' and '$\Diamond$' as operators on sentences, but in ordinary speech many different kinds of thing are said to be possible or necessary, including facts, states of affairs, and propositions, and no harm will come of exploiting this variety of means of expression. Note that '$\Box$' and '$\Diamond$' are *syntactically* just like negation; where, but only where, it is grammatically permissible to have '$-$',

[1] The terminology I take from Plantinga [1974].

[2] The philosophical *locus classicus* is Chapter 9 of Aristotle's *De Interpretatione*; see, for instance, Aristotle [1928] and the papers and bibliography in Moravcsik [1968]. A well-known modern discussion is Chapter 6 of Taylor [1974].

it is permissible to have '□' or '◇'; thus, for instance, just as 'P – Q' is nonsense, so are 'P □ Q' and 'P ◇ Q'.

2. *Invalid arguments; semantics for S5*

What *logical* principles do '□' and '◇' obey? In elementary propositional calculus, the logical principles which the operators obey are fixed by truth tables, which are said to give the *meanings* of the operators; they do this by giving the conditions under which longer sentences (formed from shorter ones using the operator in question) have particular truth values, the truth value of the longer sentence being wholly determined by the truth values of the shorter ones which make it up. We can then discover whether or not a particular argument is *valid*: for example, we accept the principle to infer 'Q' from 'P' and 'P → Q' because the truth-table for '→' tells us there are no circumstances in which 'P' and 'P → Q' are true but 'Q' is false. What we would like, then, is something analogous for '□' and '◇', so that we can answer such questions as whether it is valid to infer, say, '◇Q' from '□P' and '◇(P → Q)'. But there is no chance of providing a truth-table for '◇', since this operator lacks a property which must be possessed by any word whose meaning can be encapsulated in a truth table. Consider the table for negation:

A	–A
T	F
F	T

This tells us all we need to know about negation for the purposes of logic: a negation of a sentence takes the opposite value from the sentence negated. Suppose we try something similar for '◇'. We can certainly fill in the first row of the table:

A	◇A
T	T
F	

For if A is true 'it is possible that A' must also be true: there are ways things could have gone as a result of which A would be true, for A *is* true, and therefore the way things have actually gone is one. But we cannot fill in the second row of the table, because the mere fact that

A is false does not determine whether it *might* have been true. Suppose A is only *contingently* false; for instance, suppose A is the sentence 'Jones is a millionaire'. If we are not fatalists, we agree that Jones could have been a millionaire, so we put ◇A true. But if A is the sentence 'Jones is a married bachelor' then A is not only false, it is *necessarily* false, and so we must put ◇A false. Incidentally, it would be a mistake to dispute this example on the grounds that 'bachelor' might not have meant 'unmarried man': when we assess the truth value of *any* sentence, whether or not it contains modal operators, we take the words in the sentence to mean what they do mean, not something else.[3]

The difference which has emerged between '—' and '◇' is as follows. In order to compute the truth value of —A, it suffices to know the truth value of A. But to compute the truth value of ◇A, if the notion of computation is at all applicable, we may need more than A's truth value, for if A is false it matters whether it is necessarily or only contingently false. When a connective forms a longer sentence out of one or more shorter sentences and to compute the truth value of the longer sentence it suffices just to be given the value(s) of the shorter sentence(s), the connective is said to be *truth functional*, and it has a truth-table which gives the value of any longer sentence in terms of the values of the shorter constituent sentence(s). But when information just about the truth values of shorter sentences does not suffice to determine the values of the longer sentences the connective forms from the shorter ones, the connective is said to be non-truth-functional, and there will be at least one row of an attempted truth-table for it where we are in a quandary over what entry to make, as we were at the bottom row of the attempted table for '◇'. The reader should confirm for himself that '□' suffers from a similar problem, this time on the upper row of the table.[4]

We can make some headway with the task of providing a semantics for '□' and '◇' by considering sample incorrect inferences and asking

[3] Another way of making this point is to say that, strictly, it is not the sentence 'Jones is a married bachelor' which is necessarily false, but rather the proposition it expresses. To imagine a situation in which the sentence expresses a different proposition is irrelevant to the question of whether or not the proposition it actually expresses is necessarily false.

[4] The reader may wonder whether the impossibility of giving a truth-table for '□' or '◇' is somehow related to the fact that we are working only with two truth-values. That this is not so is shown in Dugundji [1940], where it is established that no finitary truth-tables can be given for the modal operators. However, an interesting partial characterization of the sense of the operators is possible with four values. See Kearns [1981].

why, at the intuitive level, we reject them. Consider

$$\text{(A)} \qquad \frac{\Diamond P \quad \Diamond Q}{\Diamond(P \,\&\, Q)}$$

The premisses tell us that P is possible and Q is possible, the conclusion asserts that P and Q are *com*possible. But this does not follow; it is possible that it now be raining everywhere and possible that it now be dry everywhere, but it is evidently not possible to have both these states of affairs obtaining together. A very natural way of explaining what is going on here—we shall return to the question of exactly how it is an explanation—involves yet another way of reading the operator '◇', in which a sentence of the form ◇A is read as 'there are some possible circumstances in which A'. On this reading, '◇' has become a kind of existential quantifier ranging over objects which we are calling possible circumstances. Now it is well known that the following is an incorrect inference in ordinary first-order logic:

$$\text{(B)} \qquad \frac{(\exists x)Fx \quad (\exists x)Gx}{(\exists x)(Fx \,\&\, Gx)}$$

so the assimilation of '◇' to an existential quantifier, one restricted to ranging over possible circumstances, enables us to explain the incorrectness of (A) by the familiar two-object model which explains the incorrectness of (B).

In more detail, let us use 'w', 'u' and 'v' as variables ranging over possible circumstances, and for any such variable x, let us abbreviate 'P holds in x' by 'P(x)', and 'P & Q holds in X' by 'P(x) & Q(x)'.[5] Then with the new reading of '◇' as an existential quantifier, we can re-write the argument (A) as

$$\text{(C)} \qquad \frac{(\exists w)P(w) \quad (\exists w)Q(w)}{(\exists w)(P(w) \,\&\, Q(w))}$$

The simplest formal counterexample to (B) involves a domain of two objects, one of which is F but not G while the other is G but not F, so that in this domain both premisses are true but the conclusion false (no one object in the domain is both F and G). The counterexample to (C) and therefore to (A) is just the same, except that we use different terminology to describe it. We have a domain W of two possible

[5] The reader will find a complete account of the method of translating formulae of modal propositional calculus into possible worlds language in the Appendix.

circumstances u and v. We let P hold at u but not at v and Q hold at v but not at u; so the premisses of (C) come out true while its conclusion is false; and since we are saying that '$(\exists w)P(w)$' means the same as '$\Diamond P$', '$(\exists w)Q(w)$' means the same as '$\Diamond Q$' and '$(\exists w)(P(w) \mathbin{\&} Q(w))$' means the same as '$\Diamond(P \mathbin{\&} Q)$', it follows that the counterexample to (C) is also a counterexample to (A).

If '$\Diamond$' is to be read as an existential quantifier over possible circumstances, how should '$\Box$' be read? From the equivalence of '$\forall$' with '$-\exists-$' and the definition of '$\Box$' as '$-\Diamond-$', we have little choice but to read '$\Box$' as a *universal* quantifier over possible circumstances. Furthermore, this is intuitively correct: what else could be involved in asserting that a proposition is necessary than that it holds in all possible circumstances? Now consider the inference

$$\text{(D)} \qquad \frac{P \qquad \Box(P \rightarrow Q)}{\Box Q}$$

It is easy to think of an informal counterexample. Let P be 'Jones is a bachelor' and Q be 'Jones is unmarried'. Suppose P is true; '$\Box(P \rightarrow Q)$' is also true, of course, but the conclusion '$\Box Q$' is false, for it is not *necessary* that Jones is unmarried—there are many ways things could go or could have gone in which Jones gets married. However, we can give a more formal demonstration of the incorrectness of (D) in possible circumstance terminology. '$\Box(P \rightarrow Q)$' becomes 'in all possible circumstances w, if P holds in w then Q holds in w', or in symbols, '$(\forall w)(P(w) \rightarrow Q(w))$'. What becomes of P? All of the other sentences we have considered up to now have had a modal operator as their main connective; in their possible circumstance translations, the operator becomes a quantifier and the sentential letters become *predicates* attached to the circumstance variables; if you like, the states of affairs for which P, Q, etc. stand, become properties of the circumstances. When a sentential letter stands on its own, or in a truth-functional combination with another formula, it is interpreted as making an assertion about the *actual* circumstances, which we denote conventionally by 'w^*'. So a sentential letter on its own still becomes a predicate, but a predicate of the actual circumstances w^*. Thus we translate P as '$P(w^*)$', a simple subject-predicate sentence of first-order predicate calculus. The argument (D), translated into possible circumstance terminology, now looks like this:

$$\text{(E)} \qquad \frac{P(w^*) \qquad (\forall w)(P(w) \rightarrow Q(w))}{(\forall w)Q(w)}$$

This is a straightforwardly invalid sequent of predicate calculus; consider a domain of two objects, one of which is both P and Q (this is w*) and the other neither; then the premisses of (E) are true but the conclusion false. The terminologically appropriate way of describing this counterexample is as follows. Choose a set W of two possible circumstances u and v; let each of P and Q be true at u and false at v, and let 'w*' denote u. Then P and '$\Box(P \rightarrow Q)$' are both true in the model, since true at w*, while '$\Box Q$' is false at w*, since Q is false at v. Hence (D) is incorrect. The reader should test his understanding of this argument by establishing in a similar way that:

$$\Diamond P, \Diamond(P \rightarrow Q) \models \Diamond Q$$

is also incorrect.

At this point, we will digress a little to return to our formulation of fatalism in §1. The fatalist there was represented as having the thought 'if it is going to rain tomorrow then it must be that it is going to rain tomorrow'. If we put P for 'it is going to rain tomorrow', this thought appears to have the form

(7) $P \rightarrow \Box P$

and the whole fatalist argument can be written as

$$\text{(F)} \qquad \frac{P(w^*) \qquad P(w^*) \rightarrow (\forall w)P(w)}{(\forall w)P(w)}$$

which is obviously valid. But why should we believe (7)? There may be some justice in the suspicion that the superficial attractiveness of fatalism is rooted in a failure to distinguish (7) from the triviality

(8) $\Box(P \rightarrow P)$

since by failing to distinguish (7) and (8) the incontestable correctness of (8) carries over in the mind to lend (7) a degree of plausibility it does not deserve. But if the fatalist is confusing (7) and (8) then there is no argument for '$\Box P$', since instead of (F), his argument becomes

$$\text{(G)} \qquad \frac{P(w^*) \qquad (\forall w)(P(w) \rightarrow P(w))}{(\forall w)P(w)}$$

which we have seen to be incorrect, since it is just an instance of (E). This example shows that modal logic has the same kind of clarificatory power as standard logic in the exposing of fallacies.

In establishing the incorrectness of these arguments our procedure has been to translate them into first-order predicate calculus, then to show the invalidity of the translated argument using standard semantics for first-order logic, and finally to redescribe the resulting counter-example in possible circumstance terminology. It ought to be clear that the two intermediate steps in this process are unnecessary: we should be able to go straight from the modal logical argument to the structure with possible circumstances which shows it to be incorrect. We can describe how to do this by analogy with predicate calculus. In predicate calculus, a structure or model is a domain D of objects together with a stipulation of which predicates apply to which objects in D, and of which names stand for which objects in D. Formulae are evaluated in structures by rather obvious rules which tell us, for example, that the conjunction of two predicates applies to an object iff both conjuncts do, or that an existential quantification (for instance, '(∃x)Fx') of a predicate with one free variable is true in a structure iff the quantified predicate ('F') applies to some object in the domain D of the structure. More or less the same ideas, although clothed in different terminology, give us a model theory for sentential modal logic.

In stating this model theory, we will change our earlier practice in one respect: instead of speaking of possible circumstances, we will speak of possible *worlds.* A possible world is a *complete* way things might have been—a total alternative history. The possible circumstance in which Jones takes his broker's advice and makes a million is really just a component of a world, indeed a component of many worlds, since there are many distinct total histories alternative to that of the actual world of which this circumstance is a part. In terms of our model theory, the requirement that worlds be complete is reflected in the constraint that *every* sentence letter occurring in the argument in question be assigned one or other truth value at each world. We shall see in §4 of this chapter that we can get by without this sort of completeness, but that we pay a price in terms of simplicity.

We now give the following definition of what we shall rather mysteriously call an *S5 model*:

An S5 model for a set X of sentences of sentential modal logic is

(i) a non-empty set W of possible worlds;
(ii) for each world w in W and each sentential letter S which occurs in any sentence in X, a specification of which truth value S takes at w;

(iii) a selection of a particular world w in W to play the role of the actual world; we denote this world by 'w*'.

In the terms of our analogy with predicate calculus structures, W is like the domain D of objects in such a structure, and the specification (ii) of letter-values at worlds is like the assignment of an extension from D to a one-place predicate letter. So the sentence letters are treated as if they expressed properties of worlds. (iii) has no predicate calculus parallel, for 'w*' is not a name in the language of the sentences in X, while, in predicate calculus, only names *from* the sentences in X are assigned referents in a model *for* the sentences in X.

We will say that an argument of sentential modal logic is *valid in S5* iff there is no S5 model in which its premisses are all true and its conclusion false. This is just the standard notion of validity common to all systems of logic in which sentences are true or false. However, in order to apply it, we evidently have to know how to evaluate sentences of sentential modal logic in S5 models. So next we must spell out the evaluation rules which tell us how to do this. *In general, a sentence σ is true in a model M iff σ is true at the actual world w* of M.* 'true at w*' is a special case of the more general notion of being true at a world, so our rules tell us when a sentence of a particular form is true at an arbitrary world w. The rules are exhaustive because they cover all the forms it is possible for a sentence in sentential modal logic to have.

(iv) a sentence letter π (e.g. 'P', 'Q', etc.) is true at a world w in a model M iff the specification for M described in a clause (ii) above says that π has the value true at w;

(v) a negation −A is true at a world w (in a model M—henceforth we omit this) iff A is not true (A is false) at w;

(vi) a disjunction A ∨ B is true at a world w iff A is true at w or B is true at w;

(vii) a conjunction A & B is true at a world w iff A is true at w and B is true at w;

(viii) a conditional A → B is true at w iff either A is false at w or B is true at w;

(ix) a possibilitate ◇A is true at w iff there is some world u in W such that A is true at u;

(x) a necessitate □A is true at w iff for all worlds u in W, A is true at u.

Instead of giving rules for translating modal formulae into first order formulae, we have given rules for evaluating modal formulae directly in

S5 models of the kind defined. Any formula may now be evaluated in a model with these rules, applying the rules one after another as the structure of the formula dictates. Note that in (ix) and (x), the quantifier readings of the modal operators are preserved, in virtue of the occurrence of 'there is' in (ix) and 'for all' in (x).

Now let us apply this apparatus to (A) and (D) for the purposes of illustration. To show that (A) is invalid, we need an S5 model for the set of sentences $X = \{\Diamond P, \Diamond Q, \Diamond(P \& Q)\}$ in which the first two are both true at w* but '$\Diamond(P \& Q)$' is false there. So let $W = \{u, v\}$. The sentential letters which occur in sentences in X are 'P' and 'Q', so we have to say what truth values these have at u and at v. We can do this by defining a function f for our model which assigns to each sentence letter exactly the set of worlds in W at which that letter is true. So we should say: $f(P) = \{u\}$, $f(Q) = \{v\}$. Sometimes the set of worlds at which a sentence letter is true is called the 'proposition' assigned to that sentence letter; in general, propositions are identified with sets of worlds, and the proposition expressed by any sentence, simple or complex, is the set of worlds at which that sentence is true (as determined by such rules as (iv)-(x) above). However, this is a rather special use of 'proposition'—for instance, it implies that all necessarily false sentences express the same proposition, the empty set of worlds—and we shall not employ it.

As an alternative to defining the function f, we can just draw a picture which represents the same information, and which is perhaps easier to work with:

u .	v .
P = T	P = F
Q = F	Q = T

Now let us say that w* is u; then both '$\Diamond P$' and '$\Diamond Q$' are true at w*, according to clause (ix), but according to (ix) and (vi) together, '$\Diamond(P \& Q)$' is false at w*. Hence (A)'s premisses are true but its conclusion false in this model.

For (D), X is the set of sentences $\{P, \Box(P \rightarrow Q), \Box Q\}$, and since we want P to be true in the model, it has to be true at w*. So let $W = \{u, v\}$ and put $f(P) = \{u\}$ and $f(Q) = \{u\}$. The picture is

u .	v .
P = T	P = F
Q = T	Q = F

w* must be u if P is to be true at w*, and since 'P → Q' is true at both u and v, '□(P → Q)' is true at w* as well. But by (x), '□Q' is false at w*, and so we have a counterexample to (D).

Finally, what about a *valid* argument in S5? Consider

(H) $$\frac{\Box P \quad \Box(P \rightarrow Q)}{\Box Q}$$

It is not difficult to see that there is no counterexample to (H), for if (H)'s premisses are true in a model, then 'P' and 'P → Q' are true at each world in the model, and so by *modus ponens* 'Q' must be true at each world too. The system S5 is the totality of all arguments like (H) to which there is no S5 counterexample.

3. *Other systems*

The reader will have surmised from the use of the label 'S5' that there are other systems of sentential modal logic. These arise out of divergent treatments of arguments involving sentences with iterated modalities, that is, blocks of modal operators. An example:

(I) $$\frac{\Diamond\Diamond P}{\Diamond P}$$

It is not unnatural to hear 'it is possible that it is possible that P' as making a weaker assertion than 'it is possible that P'. In support of this, one might say that there are ways things could have been such that if things had been that way it would have been possible that P, but as things *actually* are, it is not possible that P. In this case, (I) is invalid. However, (I) is valid in S5. If '◇◇P' is true at w* in some model, then '◇P' is true at some world u in the model (put '◇P' for A in clause (ix) in §2 above) and therefore P is true at some world v in the model, as in the following three-world model:

w* .	u .	v .
P = F	P = F	P = T

But then from clause (ix) it follows that '◇P' is true at w*; indeed, by repeated applications of clause (ix), we can see that if '◇ . . . ◇P' is true at any world in a model, for any number of '◇''s, then '◇P' is true at every world in the model. In terms of our intuitive argument for the invalidity of (I), the problem with S5 is that there is no way of expressing the thought that the truth of P at v does not guarantee the truth of '◇P' at w* if v is reached 'through' an intermediate world u, the idea that although P would be possible were things as in u, this does not mean that P is possible as things actually are (as they are in w*). For a counterexample to (I), we need some way of representing the idea that while what obtains at v is possible at u, it is *not* possible at w*.

What we need to do to represent this idea is, so to speak, to put barriers between worlds: if two worlds w and w′ are separated by a one-way barrier, then what obtains in w need not be possible at w′, or conversely, depending on the direction through which the barrier cannot be passed. Of course, a two-way barrier is also possible. When a world w′ is barred from a world w, so that what obtains in w′ need not be possible at w, we say that w′ is not *accessible* from w. Note that what obtains in w′ may be possible at w even if w′ is not accessible from w, since there may be other worlds which *are* accessible from w in which the same things obtain as obtain in w′. However, in the three-world model above, there is only one world, v, at which P is true, and so we can prevent '◇P' from being true at w* simply by stipulating that v is not accessible from w*. If we also say that v *is* accessible from u, then '◇P' will be true at u; therefore, if we say that u is accessible from w*, so that what obtains at u is possible at w*, then '◇◇P' is true at w*. So this gives us our counterexample to the inference (I).

We have to change some of the definitions in §2 to accommodate this new notion, accessibility. Suppose we symbolize 'y is accessible from x' as 'Acc(x, y)', which we may also read as 'x has access to y', preserving the order of the variables; then the stipulations needed to obtain our counterexample to (I) are as follows: (a) Acc(w*, u); (b) Acc(u, v); (c) Not-Acc(w*, v). It is immediately obvious from (a), (b), and (c) that what we have done to obtain the counterexample is to allow the accessibility relation to be non-transitive; that is, we have *not* insisted that:

$$(8)\quad (\forall x)(\forall y)(\forall z)(\mathrm{Acc}(x, y)\ \&\ \mathrm{Acc}(y, z) \rightarrow \mathrm{Acc}(x, z)).$$

So the example suggests that which arguments are counted as valid or invalid will turn on which structural conditions (transitivity, symmetry, etc.) we impose or refuse to impose on the accessibility relation. Let us now reformulate some definitions from §2 with all this in mind. Roughly, we want to make room for the accessibility relation in our definition of model and we want to alter (ix) and (x) so that the truth value of a modal sentence at a world w depends only on what obtains at the worlds accessible from w.

A *general* model for a set X of sentences in sentential modal logic is:

(i) a non-empty set W of possible worlds;
(ii) for each sentence letter π which occurs in any sentence in X, and for each world w in W, a specification of the truth value of π at w;
(iii) for each world w in W, a specification of which worlds in W are accessible from w;
(iv) a selection of a particular world w from W as w*.

Here clause (iii) is the novelty; note that (iii) does not impose any structural constraints at all on the accessibility relation; (iii) even admits a model in which no world is accessible from any world, not even itself; to rule out such a model, we would have to stipulate that accessibility is reflexive, i.e. that each world has access to itself.

All the rules in §2 for evaluating formulae at worlds in S5 models carry over to general models, except the rules for the modal operators. In place of (ix) and (x) from §2, we have the following:

($\Diamond$) a possibilitate $\Diamond$A is true at a world w in a general model M iff there is some world w′ in the set W of worlds in M such that A is true at w′ and w′ is accessible from w according to the specification of accessibility described in (iii) above for M;
($\Box$) a necessitate $\Box$A is true at a world w iff for all worlds w′ in W, if w′ is accessible from w then A is true at w′.

Different systems of modal logic arise as we impose different structural constraints upon the accessibility relation; moreover, relationships between these constraints will be reflected in relationships amongst the systems they generate. Thus if one set of constraints on accessibility S includes all those of some other set S′, then every model for the system generated by S will be a model for the system generated by S′, and so any invalid formula in the first of these systems will be invalid in the second, since the first system's model which refutes it is also a model for the second system. Hence the valid formulae of the second, S′-system, are valid in the first, the S-system, as well, so the second system is a subsystem of the first.

To proceed to actual examples, let us start with the constraints on accessibility which generate the system S5. That is, we want to define a subclass of general models which validates exactly the arguments validated by our earlier account of an S5 model. Truth in a model, as always, is truth at the actual world of a model, and in our original S5 models we allowed that any world in W is accessible from w*, because we did not even mention accessibility in the definition of a model or in the clauses for the modal operators. For the same reason, then, we in effect allowed that any world in W is accessible from any other world in W, and, of course, also from itself. Thus in S5 models of our original sort, every world in W is equivalent to any other world in W with respect to what worlds are accessible from it. This suggests that if we take the subclass of general models in which the accessibility relation is an *equivalence* relation (reflexive, symmetric and transitive) we obtain exactly the S5 models.

However, there is a slight complication. In a general model in which accessibility is an equivalence relation, the set of worlds W may be partitioned into a *number* of different equivalence classes, in much the same way as the equivalence relation 'is of the same nationality as' partitions the set of people into difference equivalence classes—Americans, Brazilians, etc. (suppose no one has dual nationality). Such a general model is not *identical* with any S5 model of our original sort, since in these there is only one equivalence class of worlds, the whole set W. However, it is easy to see that for any S5 model M of the original sort, there is a class C of general models in which accessibility is an equivalence relation, such that a sentence is true in M iff it is true in every model M′ in C. In each of these models M′, the equivalence class to which w* belongs is exactly like the model M—the same sentence letters are true at the same worlds.[6] In general, there are also other equivalence classes of worlds in any M′, but *it does not matter* what sentence letters are true or false at worlds in these equivalence classes: what is possible at w* depends *only* on what obtains at the worlds in the equivalence class of w*; no matter how many modal operators a sentence A may contain, in evaluating A at w*, one never goes outside the equivalence class of worlds to which w* belongs, since for each w in this class all (and only) the worlds accessible to w are also in it.

[6] This remark is not wholly precise, since the worlds in M may be different *objects* from the worlds in the equivalence class of the actual world of M′. The situation is further complicated by the fact that in the model theory nothing prevents the same assignment of truth-values to sentence letters being made at different worlds although, intuitively, these different formal objects would be representations of the same possible circumstances. So it is simpler to state the relationship relevant here in terms of these assignments: each truth-value assignment made at some world in M is made at some world in the equivalence class of the actual world of M′, and conversely.

Analogously, for each American x, the class of Americans contains exactly, hence all, those people of the same nationality as x. One may think of one of our original S5 models as sitting 'inside' various general models, those in which accessibility is an equivalence relation and from which the original S5 model could be obtained by first deleting all the equivalence classes of worlds other than the class of w*, and then dropping all mention of accessibility. Since the same sentences hold in all models thus related, it follows that an argument has an S5 counterexample of the original sort if and only if it has a counterexample involving a general model in which accessibility is an equivalence relation. So the system S5 is the system whose models are exactly those general models in which accessibility is reflexive, symmetric and transitive.

The three structural properties of accessibility characteristic of S5 suggest a natural way of obtaining other systems, by dropping one, two, or all three of the properties. We have already seen that to obtain a counterexample to the inference (I) we have to permit accessibility to be non-transitive. So if we make the three stipulations (a), (b), and (c) of five paragraphs back about the model we pictured, and then add the further stipulations (d) Acc(w*, w*), (e) Acc(u, u), (f) Acc(v, v), (g) Acc(u, w*) and (h) Acc(v, u), we obtain a general model in which accessibility is non-transitive, symmetric and reflexive. The class of all models in which the accessibility relation is reflexive and symmetric, but not necessarily transitive (it will be transitive in some models in the class but not in others), determines a set of valid inferences, usually known as the system B, because of a tenuous connection with the intuitionistic logic of Brouwer.[7] So the general models which are B models consist in all S5 models together with those other general models in which accessibility is non-transitive (though still reflexive and symmetric). Thus any S5 counterexample to an inference will also be a B counterexample, since S5 models are B models, but some B counterexamples to inferences will not be S5 counterexamples, since no B model which refutes the inference is an S5 model (these are the inferences, such as (I), refuted only by B models with non-transitive accessibility relations). So all B validities are valid in S5, while some S5 validities are invalid in B, for instance, (I). It follows that B is a subsystem of S5, for its validities are a subset of those of S5; this is a

[7] For further information about this and other aspects of the historical development of modal logic, see Hughes and Cresswell [1968 *passim*], and Lemmon and Scott [1977].

concrete instance of the kind of relationship we described above which arises between systems in virtue of the underlying relationship between the sets of constraints on accessibility which generate the systems.

Another subclass of general models is the class whose members all have reflexive and transitive accessibility relations, while in some of them accessibility is symmetric and in others non-symmetric. Consider the inference:

$$\text{(J)} \qquad \frac{P}{\Box\Diamond P}$$

For '$\Diamond$P' to be true at w*, all we require is that a world at which P is true be accessible from w*, but for '$\Box\Diamond$P' to hold at w*, every world accessible from w* must be such that P is true at some world accessible from *it*. If accessibility is not necessarily reversible (not symmetric) then this second requirement will fail even if P holds at w*. Consider the general model:

w* . ⟶ u .
P = T   P = F

In such diagrams as these, an arrow running from x to y represents the accessibility of y from x, so in this particular diagram, w* is not accessible from u, since there is no arrow running *from* u to w*. Hence, although P is true at w*, '$\Box\Diamond$P' is false there, since there is a world accessible from w*, that is, u, at which '$\Diamond$P' is false. '$\Diamond$P' is false at u because P does not hold at any world accessible from u. In this model, accessibility is non-symmetric because there is an arrow between w* and u in one direction but not in the other.

The system whose models are all and only the general models in which accessibility is reflexive and transitive is called S4, and so the model we have just described is an S4 model, since, as the arrows are drawn, it is only symmetry which fails. Thus (J) is not valid in S4. Neither S4 nor B is a subsystem of the other, since the S4 model just described is not a B model, and the B model described previously is not an S4 model and, in fact, (J) is valid in B and (I) is valid in S4. However, the S5 models are a subclass of S4 models, just as they were of B models, and so by the same reasoning as showed that B was a subsystem of S5, we can infer that S4 is also a subsystem of S5.

A fourth system results if we relax the restrictions on accessibility even further, and require only that it be reflexive. The system whose models are exactly the general models in which accessibility is reflexive is called T, and it should be clear that T is a subsystem of B, S4, and S5, since every model for any of these three systems is also a T model, but some T models, those in which accessibility is non-transitive *and* non-symmetric, are not models for any of these other systems. So we can picture the inclusion relations amongst the systems in the following figure, with S5 at the top, since it has most valid arguments, the other systems all included in it.

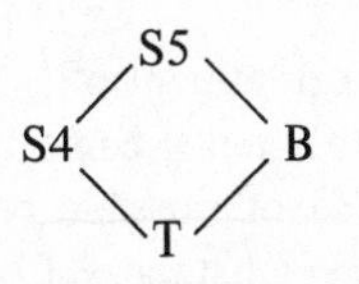

Turning the diagram upside down yields a representation of the inclusion relationships amongst the classes of models for the systems.

A fifth system which we can generate by these methods is the system whose models are all the general models, the system for which there are no restrictions on accessibility. But this system does not appear to capture any natural notion of possibility or necessity. For instance, it would be incorrect in this system to infer '◇P' from P: consider a model with one world, w*, at which P is true, in which w* is not accessible from itself (accessibility is non-reflexive). Then '◇P' is false at w*, so we have a counterexample to the inference. It is also possible (reader's exercise) to give a counterexample to the inference of P from '□P'. Such a system is of little interest in the present context.[8]

Clearly, we have not exhausted the different possible stipulations which we can impose on accessibility: we could demand that any world be accessible only from a finite number of worlds, that of any two worlds at least one be accessible from the other, and so on. Such stipulations select different subclasses from the class of general models and so determine different systems: it is hardly possible to give a complete account here.[9] But with such a plurality of systems, it is natural to ask which is the 'right' one. For our purposes, the right one is the one which captures our concepts of broadly logical possibility and necessity. We have said enough about these concepts to see that they are not captured by a system which permits models with non-reflexive

[8] But it is of considerable interest in a context in which '□' stands for provability in formal arithmetic, rather than broadly logical necessity. See Boolos [1979].

[9] A very full account is provided by Chellas [1980].

accessibility, but we have not said enough to make further discriminations. However, it will turn out to be S5 which we want, and this foresight motivates restricting further development of modal semantics to S5 alone, which means that we can drop mention of accessibility; an S5 model can be defined as it was in §2.[10]

4. *Incomplete circumstances: possibility semantics*[11]

In motivating the quantifier treatment of '□' and '◇', we used the phrase 'possible circumstance', but when it came to defining the notion of an S5 model, we spoke instead of possible worlds, with the explanation that worlds are *complete* circumstances. Completeness amounts to the condition that in each world in a model every sentence letter in the language of the relevant argument is assigned a truth-value, the language of the argument being the set of all sentence letters that occur in any sentence in the argument. It is not evident in the resulting semantics what role this completeness condition plays, and the simplest way of highlighting the role is to see what we have to do to get by without completeness. So our question is: if we allow worlds to be incomplete, how can we obtain the system S5?

Let us say that each sentence of the form '◇A', where A itself contains no modal operators, introduces a certain *possibility*: obviously

[10] The reader who has absorbed the fundamental ideas of the formal workings of accessibility is in a position to raise a philosophical query about its motivation. The argument used in the text to introduce the relation is that one can hear '◇◇P' as making a weaker assertion than '◇P', since P could be possibly possible but not possible as things actually are. I must emphatically state that I am not endorsing this claim about how one can hear the two assertions. Someone who thinks that the first is weaker than the second may well be equivocating on 'possibly' where it is iterated. For instance, it is uncontroversially true that a proposition may fail to be *causally* possible as things actually are, i.e., given the actual laws of nature, but it may be possible in the broadly logical sense that the proposition is possible in the causal sense: the idea would be that in some logically possible world the laws of nature are different, and the proposition in question does not contradict *those* laws of nature, even if it does the actual laws. But when we are careful to mean the broadly logical sense both times in 'possibly possibly P', it is very unclear that a difference in sense between this assertion and 'possibly P' can be grasped. In fact, it is unclear that the accessibility relation corresponds to any components of our thinking about broadly logical possibility. If this is right, only S5 is of interest in the present context. See further note 26 to Chapter 9.

[11] This section is based upon Humberstone [1981], and is rather more advanced than the preceding sections. The reader may prefer to proceed directly to Chapter 2, and read through to the end of Chapter 3, omitting §4 of Chapter 2, which extends Humberstone's semantics to first order logic. It is not until the end of Chapter 4 that the results of these two sections are appealed to.

enough, the possibility that A. One can think of the possibility that A as a possibility which can be realized in a number of different ways, corresponding to the different assignments of truth-values to the sentence letters in A on which A is true. By analogy with worlds, then, one can think of a possibility as an object at which certain sentence letters are assigned truth values, and on some of these assignments, one can say that A is true at that possibility. It is clear that possibilities will be incomplete in our sense, for '◇A' may be just one premiss in an argument whose other premisses, or conclusion, contain sentence letters which do not occur in A. So A could be true at a possibility even although there are some sentence letters in the language which are not assigned any truth value at that possibility. The idea, then, is that possibilities are a broader class of entities than possible worlds: they include worlds, which are possibilities which happen to be complete, but also include 'incomplete worlds', a phrase which strictly speaking is self-contradictory: worlds are complete by definition, possibilities need not be.

According to this rough account, completeness is a relative notion: an entity is complete *with respect to a language* L iff each sentence letter in L is assigned one or other truth value at that entity. So in the discussion to follow, the relativity to language will be explicit. We now want to define the idea of a possibility model and give evaluation clauses for the connectives in such a way that possibility models validate exactly the arguments validated by the S5 models of §2 (a more general treatment which allows for an accessibility relation and yields something analogous to the notion of a general model is also possible[12]).

The initial steps are quite clear. Part of the definition should go like this:

A possibility model for L ('p-model for L') is

(i) a set P of possibilities;
(ii) for each possibility X in P, and for some subset of the sentence letters of L, a specification of which truth value each sentence letter in the subset has at X;
(iii) a selection of a particular possibility X to play the role of the actual world p*; since the actual world is, at least arguably, complete, X should be a possibility at which every sentence letter in L is assigned a truth-value.

[12] See Humberstone [1981].

However, we need a little more apparatus if we want to end up with an S5 semantics. Consider two possibilities X and Y in a p-model for L, and suppose that the sentence letters which are assigned a truth value at X are a subset of those assigned a truth value at Y. Then if every sentence letter assigned a truth value at X retains that value at Y, we say that Y is a *refinement* of X, or that Y *refines* X. We will symbolize this relationship with '$\geqslant$', so if Y refines X we write '$Y \geqslant X$'. Let us also say that the sentence letters which are assigned some value at a possibility X are the sentence letters for which X is *defined*. Then '$Y \geqslant X$' means that Y assigns the same truth-values as X does to the sentence letters for which X is defined. We can now state the extra clause we need in the definition of p-model:

(iv) *Refinability*: for any sentence letter π in L and any possibility X in P, if π is undefined (neither true nor false) at X, then there are possibilities Y and Z in P such that $Y \geqslant X$ and $Z \geqslant X$ and S is true at Y and S is false at Z.

The technical justification for clause (iv) is just that without it we do not even obtain classical propositional logic, much less S5. Intuitively, the clause says that every possibility may be further refined, or extended, in each of two ways, for any atomic proposition undefined at that possibility. We might say that this is guaranteed by the mutual logical independence of the atomic propositions, and if it is possible that a possibility be refined in such-and-such a way, then *there is* a possibility which is the refinement of the original one in that way. So Refinability is a rather natural closure condition on the domain of a possibility model.

The next step is to give evaluation clauses for each of the connectives, and it is here that there is a considerable loss of simplicity by comparison with possible worlds semantics. Two clauses are rather obvious, corresponding to clauses (iv) and (v) in §2.

(v) a sentence letter π is true at a possibility X in a p-model M iff the truth value specification described in clause (ii) assigns the value T to π at X;

(vi) a conjunction A & B is true at a possibility X iff A is true at X and B is true at X.

All we need to do now is to give a clause for negation, and then we can treat all the other connectives as being introduced by the usual definitions in terms of '—' and '&'. But under what condition should a negated sentence —A be said to be true at a possibility X? Suppose that A is the sentence letter 'Q'. If 'Q' is false at X, then '—Q' should certainly be true there, and vice versa. But if 'Q' is undefined at X, then it verges on inconsistency with Refinability that '—Q' should be true. According to Refinability, some possibility Y refines X and 'Q' is true at Y; if '—Q' is true at X, it follows that although the refinement relation preserves the truth values of atomic sentences, it does not preserve the values of complex sentences: otherwise, 'Q' and '—Q' would both be true at Y. But this clashes with our conception of refinement as elaboration or expansion of a possibility rather than revision of it. For non-atomic sentences, however, the problem will not arise if A is undefined at X and there is no refinement of X at which A is true. This prompts the thought that for any sentence A, the condition under which —A should be true at X is just if A is not true at any refinement of X. The clause is:

(vii) —A is true at X iff there is no Y such that $Y \geqslant X$ and A is true at Y.

It is now possible to introduce the other propositional connectives by definitions, and it is worth investigating what evaluation conditions these definitions bestow on the connectives. For instance, by using (vi) and (vii) to unpack '—(—A & —B)', we find that

(viii) A v B is true at X iff for every Y which refines X there is some Z which refines Y such that A is true at Z or B is true at Z.

The inconvenience of working with incomplete worlds is apparent from this clause. However, the clauses for the modal operators are simple. For '◇', we just say

(ix) ◇A is true at X iff there is some Y such that A is true at Y.

The connective '□' can now be introduced by the definition '—◇—', and, again by unpacking, it can be established that the evaluation condition for □A thus bestowed is the standard one, requiring the truth of A at every possibility.

Next, we have to show that the semantics just outlined validates the same arguments as the S5 possible worlds semantics of §2. For the details, the reader must refer to Humberstone [1981], but the main idea is not difficult. To show that the two semantics validate the same arguments, it suffices to show that for any set Σ of L-sentences, Σ has an S5 possible worlds model iff Σ has a possibility model as defined above. One half of this biconditional is immediate, for if Σ has an S5 possible worlds model, it has a possibility model *ipso facto*, one in which the possibilities are all complete and the only refinement any possibility has is itself (it was to permit this move that we said that possibilities may be, but do not *have* to be, incomplete).[13] For the other half of the biconditional, it is necessary to take some set of axioms and/or rules of inference from which exactly the valid S5 arguments can be proved, and show that each axiom is true in any possibility model and that each rule of inference preserves truth in any possibility model. This can be done. Thus if Σ has no S5 possible worlds model, so that Σ is S5 inconsistent and we can infer an explicit contradiction from the sentences in Σ using the axioms and rules for S5, the soundness of these axioms and rules with respect to possibility models means that Σ has no possibility model. Hence, if Σ has a possibility model, it must have an S5 model. Thus the semantics of §2 and the possibility semantics validate the same arguments.

Despite this equivalence, we shall not make much use of possibility semantics: the clause for negation makes the system too complicated to work with, especially when we move to first-order modal logic. However, for certain philosophical purposes, it will be useful to know that it is in principle possible to employ incomplete circumstances; we will return to this topic in Chapter 5.

[13] It is also possible to give a possibility semantics for S5 in which complete possibilities are disallowed. See Humberstone [1981, p. 326].

2
First Order Modal Logic

1. *Operators, quantifiers and invalid inferences*

JUST as sentential modal logic was obtained by adding '□' and '◇' to ordinary sentential logic and allowing these operators to function syntactically like '—', so modal predicate calculus is obtained by adding '□' and '◇' to ordinary predicate calculus, first order logic with identity, where they are again governed by the same syntactic formation rules as govern '—'. So we can have '□(Fa)' and '◇(Fa)', but not 'F □ a' and 'F ◇ a', since we cannot have 'F — a'.

Intuitions about what semantics for modal predicate calculus should be like are best engaged by consideration of how modal operators interact with the characteristic connectives of predicate calculus, the existential and universal quantifiers. The reader is familiar with the distinction between '$(\exists x)-Fx$' and '$-(\exists x)Fx$', and between '$(\forall x)-Fx$' and '$-(\forall x)Fx$'. We begin, therefore, by investigating the question of whether there are corresponding differences between such pairs of formulae as '$(\exists x)\Diamond Fx$' ('something is possibly F') and '$\Diamond(\exists x)Fx$' ('possibly something is F'), and '$(\forall x)\Box Fx$' ('everything is necessarily F') and '$\Box(\forall x)Fx$' ('necessarily, everything is F').

An atheist is someone who holds that there is no such being as God, but he may think of this as either a contingent or a necessary truth. Consider the atheist who believes that God's non-existence is contingent, perhaps because he thinks that the mere idea of a being like the God of traditional Christianity is not in itself unintelligible. As there are certain difficulties with proper names for entities which do not exist—indeed, the phrase 'entity which does not exist' is *prima facie* self-contradictory—this atheist's view is more properly expressed as the claim that there could have been such a being as God, where the phrase 'such a being as God' is elliptical for a list of predicates which are supposed to characterize this being: 'omniscient', 'omnipotent', etc. Let us summarize this list in the predicate 'God-like'. Then the first part of our atheist's view, that there is no such being as God, may be formalized in this way:

(1) $-(\exists x)$(x is God-like).

But how should we formulate the second part of his view, that there could have been such a being? There are evidently two candidates:

(2) $\Diamond(\exists x)$(x is God-like)

and

(3) $(\exists x)\Diamond$(x is God-like).

English renderings of (2) and (3) which respect the relative orders of the modal operator and quantifier in them are, respectively,

(4) It could have been that there is a God-like being

and

(5) There is a being which could have been God-like.

It is obvious that (4) and (5) do not mean the same, but unclear exactly what the difference between them is. However, we can bring the difference out by applying to (4) and (5) the quantifier treatment of the modal operators we developed in the previous chapter. It will be recalled that in translating such a sentence as '$\Diamond$P' into possible worlds language, the sentence letter 'P' became a one-place predicate of worlds, to yield the formula '$(\exists w)Pw$'. We may think of a sentence letter as a 0-place predicate letter, so the generalization of the propositional calculus translation will require n-place predicates to become n + 1-place predicates; for instance, the predicate 'Rx' ('x is red') will become 'Rxw' ('x is red at w').[1] From (4) and (5) we then obtain:

(6) There is some world w such that there is some object x which is God-like at w ('$(\exists w)(\exists x)Gxw$');

and

(7) There is some object x such that there is some world w such that x is God-like at w ('$(\exists x)(\exists w)Gxw$').

Let us now ask what is required of the actual world, w*, if (6) and (7) are to be true there. (7) requires that there is some object x which meets a certain condition, and if it is true at w* that there *is* an x

[1] A rigorous translation scheme is given for first-order modal logic in the Appendix.

meeting this condition (that of being God-like in some world) then that object must exist at w*; otherwise, it cannot be true at w* that *there is* such an object. This means that (7), and therefore (5) and (3), express the view that something which actually exists could have been God-like, even although it actually is not, if atheism is true. It is unlikely, then, that (3) is what our atheist means when he says that God's non-existence is contingent, for it is unlikely that he holds that some actual object could have been just as God is supposed to be. Perhaps he believes that every actual object is made of matter, and it is not very easy to imagine a way things could have gone in which something which is actually material turns up as a 'spiritual' being.

So it looks as if (2), (4), and (6) are the formulations which we seek. (6) requires of the actual world that there be some world in which some being is God-like. Let us suppose that there is such a world, say, u. Then 'there is some God-like being' is true at u, which seems to require that in u, a God-like being *exists*; otherwise, it would not be true at u that *there is* such a being. It may be wondered if it is consistent to hold both that no actual object is God-like at any world, and that at some world, some being is God-like. But there is really no difficulty here. All we have to assume is that *there could have been things which do not actually exist*, and then we can think of the God-like being in u as one of these things. Indeed, it is natural to suppose that this is what our atheist has in mind. In sum, we can now see that the difference between (6) and (7) amounts to a difference in the 'domain' ranged over by the existentially quantified variable 'x': in (7) it ranges over the actual world, but in (6) it ranges over whatever world is introduced by an evaluation of the existential quantifier over worlds. We will make this more precise below.

We can bring out the difference between '$\Box(\forall x)Fx$' and '$(\forall x)\Box Fx$' with a similar example. The atheist we are imagining believes that everything is made of matter, a belief which is true at the actual world iff all actually existing things are made of matter; therefore, his belief is consistent with the view that there could have been things which are not made of matter. But let us imagine a different atheist who believes that it is a necessary truth that everything is made of matter; this atheist denies that there could have been a God-like being, since immateriality is one of the supposed attributes of God. What is the correct modal formulation of his view about matter? Again, there are two candidates:

(8) $\Box(\forall x)$(x is made of matter)

and

(9) $(\forall x)\Box$(x is made of matter).

(8) would seem to be correct. Reading '$\Box$' as a universal quantifier over worlds, (8) says that at every world w, '$(\forall x)$(x is made of matter)' is true, and this demands of each world w that everything in w, that is, everything which exists at w, be made of matter in w (concomitantly with our treatment of the objectual quantifier in (6), the quantifier in (8) ranges over only the existents of a world). So (8) says that there are no worlds in which there exist non-material things. (9) appears to be doubly inappropriate as a formulation of *a priori* materialism. If (9) is true at the actual world, then for every actual object x, it is to be necessary that x is made of matter, that is, every such x is to be made of matter at every world w. So if (9) is true at the actual world, then each actual object is made of matter in every possible world. Presumably, however, an object cannot be made of matter at a world unless it exists at that world. So (9) implies that every actual object exists at every world, which conflicts with the common sense view that material things exist contingently. We can alter (9) so as to eliminate this unwanted implication:

(10) $(\forall x)\Box$(Exists(x) $\rightarrow$ x is made of matter).

(10) says that for each object x, for each world w, if x exists at w then x is made of matter at w, that is, x is made of matter at every world *at which x exists*. This is quite consistent with x's failing to exist in any number of worlds, but the truth of (10) at the actual world is still not sufficient to express our second atheist's view; for (10)'s being true at the actual world is consistent with there being some world in which there is something which is not made of matter, something which, *a fortiori*, does not actually exist. And this is inconsistent with the materialist doctrine expressed by (8).

With these examples before us, we can reiterate the main points more formally. Let 'E(x, w)' be a predicate which means 'x exists in w', and let 'Trans[A]' stand for the result of translating a formula of modal predicate calculus into possible worlds language according to the principles employed above. Then (the translation scheme in §2 of the Appendix gives us):

(11) $\text{Trans}[\Diamond(\exists x)Fx] = (\exists w)(\exists x)(E(x, w) \,\&\, Fxw))$;

(12) $\text{Trans}[(\exists x)\Diamond Fx] = (\exists x)(E(x, w^*) \,\&\, (\exists w)(Fxw))$;

(13) $\text{Trans}[\Box(\forall x)Fx] = (\forall w)(\forall x)(E(x, w) \rightarrow Fxw);$

and

(14) $\text{Trans}[(\forall x)\Box Fx] = (\forall x)(E(x, w^*) \rightarrow (\forall w)Fxw).$

(11) should be compared with (2), (4), and (6), and (12) with (3), (5), and (7). Note that these translations use the existence predicate to make explicit our intuitive judgements about the meanings of (6) and (7), according to which the variable 'x' ranges over different domains (worlds) in the two formulae. (13) and (14) differ similarly; according to (13), 'Necessarily, everything is F' means that in every world w, everything which exists at w is F at w, while according to (14), 'everything is necessarily F' means that everything which actually exists is F at every world. By the same principles, if we translate a formula with the form of (10), we obtain

(15) $(\forall x)(E(x, w^*) \rightarrow (\forall w)(E(x, w) \rightarrow Fxw)),$

that is, 'every actual object is F at every world at which it exists'.

By inspecting their translations, we can see that (instances of) the following schemata are not logically true.

(16) $\Diamond(\exists x)Fx \rightarrow (\exists x)\Diamond Fx$

is not logically true, essentially because worlds contain objects which do not actually exist, so that for particular instances of 'F', e.g. 'Godlike' (on the first atheist's view) the conditional has a true antecedent and a false consequent. Similarly, the conditional

(17) $\Box(\forall x)Fx \rightarrow (\forall x)\Box Fx$

is not logically true: the antecedent might be true at the actual world since everything is F in every world, but the consequent false at the actual world, since not every actual object exists at every world (a clearly false substitution instance is obtained if 'Exists' is put for F). Furthermore, elementary manipulations of (16) and (17), exploiting the interdefinabilities of '$\Box$' and '$\Diamond$', reveal that (16) is logically equivalent to

(18) $(\forall x)\Box Fx \rightarrow \Box(\forall x)Fx$

and (17) is logically equivalent to

(19) $(\exists x)\Diamond Fx \rightarrow \Diamond(\exists x)Fx$

which therefore also fail to be logically true. To see directly that (19) is not logically true, put 'does not exist' for 'F'. Then the antecedent of the resulting instance of (19) is true at the actual world if some actual thing might not have existed, but the consequent is a *contradictio in adjecto*, since it requires that there be some world at which *there is* something which does not exist. In the literature on modal logic, (16) and (18) are sometimes referred to as 'Barcan' formulae, and (17) and (19) as 'converse Barcan' formulae.[2]

2. *Semantics for quantified S5*

The discussion of the previous section should have imparted a general picture of what model theory for quantified S5 is going to look like. As in the sentential case, there will be a set of possible worlds, but in addition, each world will be assigned a set of objects, the things which exist at that world. The sum total of all objects which exist at the various worlds forms the set of all possible objects, and the atomic predicates and relation symbols of the language will be assigned extensions at each world drawn from this universal set, just as sentence letters were assigned truth values at each world. The precise details were given a classic formulation by Saul Kripke, and the presentation below is essentially a version of Kripke's semantics due to Kit Fine.[3]

We say that an S5 model M for a set X of sentences of quantified modal logic consists of the following six components:

(i) a non-empty set W of possible worlds;
(ii) a non-empty set D of possible objects;
(iii) a function d which assigns to each world w in W a subset d(w) of D; d(w) is the set of objects which exist at w, sometimes called the domain or inner domain of w; the function d satisfies the condition that for every x in D there is some w in W such that x is in d(w) (every possible object exists in at least one world);
(iv) for every n-place atomic predicate F occurring in some sentence in X and for each world w, a specification of which n-tuples of objects drawn from D (NOT: 'drawn from d(w)') are in the extension of F at w;
(v) for each individual constant occurring in some sentence in X, an assignment to it of a referent, some object in D (so the reference of a constant is the same at every world);

[2] After the logician Ruth Barcan Marcus. See Marcus [1962].
[3] See Kripke [1963] and Fine [1978a].

(vi) a selection of a particular w in W as w*.

As before, a sentence is true in a model M iff it is true at w* in M. Quantified S5 is the system whose validities are exactly those arguments validated by the class of models conforming to (i)–(vi). These clauses define a notion of S5 model which is a natural extension of the notion defined in §2 of Chapter 1. In that sentential case, each possible world was associated with a model for non-modal sentential logic, that is, an assignment of truth values to sentence letters. Here we are associating each world with something quite like a model for non-modal predicate calculus, which, as the reader knows, consists in a domain of discourse D and an assignment from D of extensions to predicate letters and referents to names. But because of clauses (iv) and (v), the parallel is not quite exact, and some further comments on these two clauses are in order.

Clause (v), according to which the referent of a constant at w is fixed independently of the nature of w and need not occur in d(w), is quite natural; for example, if the sentence 'Socrates does not exist' is to be true at a world w where Socrates does not exist, then if this sentence is to be treated on a par with any other sentence of the form '—F(Socrates)', its truth at w must consist in the referent of 'Socrates' at w not being a member of the extension of 'exists' at w (which is d(w)); and because the sentence has to be true at w in virtue of its being *Socrates* who does not exist at w, it follows that the referent of 'Socrates' at w must be Socrates, even though he does not exist at w. In sum, the natural treatment of 'Socrates does not exist' demands that the name 'Socrates' denote Socrates at all worlds, even those at which Socrates does not exist. A constant which designates the same object at every possible world is sometimes called a *rigid designator*. It is a somewhat controversial thesis that the proper names of natural language are rigid designators, though it seems to be borne out by our example, but the controversy is irrelevant to the concerns of this book.[4]

Clause (iv) allows an atomic predicate to hold at a world w of an

[4] The standard reference is Kripke [1972, pp. 278–303]. For objections, see Schiffer [1978]. The reader familiar with the literature may be surprised at the claim that the controversy about names is irrelevant to the metaphysics of modality, since arguments about the necessity of identity appear to turn on how names are treated. But this is a misconception, as is explained in §5 of Chapter 3; the necessity of identity asserts that one thing could not have been many, nor many things one, a thesis with no implications for the semantics of proper names.

object which does not exist at w, and this seems strange, because the predicates of English which one thinks of as the natural candidates for regimentation by the atomic predicate letters of a formal language are usually predicates which can be satisfied only by existents. The recently noted example 'made of matter' is a case in point. Following Fine, we may explain this as arising from the fact that properties and relations which are in some sense 'genuine' hold at any world only amongst the existents of that world, and we reserve regimentation by the atomic predicates of a formal language for expressions, such as 'made of matter', which pick out these genuine properties and relations. When we have an expression, such as 'does not exist', which can be true, at a world, of non-existents at that world, we are inclined to think of the property it picks out as being in some way complex, and analysable into genuine components, so we would be reluctant to regiment it as an unstructured expression. Fine distinguishes two versions of this attitude towards primitive predicates and non-existents. On one, primitive predicates expressing genuine properties and relations neither apply nor fail to apply to non-existents at a world, while on the other, they apply, but are false of them. Since the second of these views is technically easier to implement, we shall work with it, that is, we shall work with what Fine calls the Falsehood Principle, which in full generality says that at any world, no n-tuple of possible objects satisfies any n-place primitive predicate at that world unless every member of the n-tuple exists at that world.[5]

Even if we are willing to assert the Falsehood Principle outright, however, clause (iv) of the model theory would still be justifiable, since it is evident that the considerations which recommend the Principle are highly philosophical in nature, and it would be bad practice to build such a philosophical view into modal logic itself: a dispute, for example, about the content of the notion of genuineness employed above (a question about which we shall remain agnostic) should not be misrepresented as a dispute about logic. Moreover, there are even stronger grounds for retaining (iv), since not every predicate or relation which holds amongst existents and non-existents, or amongst non-existents and non-existents, stands for a complex with genuine components in the rather obvious way in which 'does not exist' does (we give an example in footnote 22 to Chapter 7). So one may want to

[5] My discussion here draws heavily on Fine [1981a, pp. 293–4]. For a version of S5 based on the idea that the atomic predicates neither apply nor fail to apply at a world to non-existents at that world, see Davies [1978].

formalize some predicates as atomic and then, as a further step, raise the question whether or not the Falsehood Principle applies to each of them. From this point of view, it is better to regard the Principle as a *schema*, with some true instances and (perhaps) some false ones. The schematic version of the Principle is:

$$\text{(F)} \quad \Box(\forall v_1) \ldots \Box(\forall v_n)\Box(Hv_1 \ldots v_n \rightarrow E(v_1) \,\&\, \ldots \,\&\, E(v_n))$$

where H takes only atomic substituends (the reader who has difficulty construing (F) at this point should return to it at the end of this section). So if we decide, in any particular context, to assert (F) of some atomic predicate, that will be to formulate a non-logical axiom governing that predicate; where 'Mx' abbreviates 'x is made of matter', we have already adopted this instance of (F):

$$\Box(\forall x)\Box(Mx \rightarrow E(x)).$$

Finally, in remarking on clauses (i) to (vi), we should note some consequences, for the set of validities, of the conditions on the function d which assigns domains to worlds. Because D is non-empty and d must assign each object in D to at least one world, the formula

$$\Diamond(\exists x)E(x)$$

which means 'possibly something exists', is a logical truth. But because nothing prevents d from assigning the empty set to some world as its domain (recall that the empty set is a subset of every set), a valid schema of standard predicate calculus,

$$(\forall x)Fx \rightarrow (\exists x)Fx$$

will fail at some worlds. In fact, in our system of S5, the logic of the objectual quantifiers is that for the empty domain.[6] Additionally, our treatment of quantifiers, on which the quantified variable ranges over just the domain of the world at which the quantified formula is being evaluated, has the consequence that '$(\forall x)E(x)$' is trivially true at every world, so that it, and also

$$\Box(\forall x)E(x)$$

[6] In such a system, universally quantified variables in a formula can be instantiated in the course of a proof only by names which occur in the premisses or negation of the conclusion. Hence, since there are no names in '$(\forall x)Fx$', and since '$-(\exists x)Fx$' becomes '$(\forall x)-Fx$', no further moves can be made and the tree remains open. A tableau method of syntactically testing for validity with such a $\forall$-rule is presented in Hodges [1977]. For further discussion, see Scott *et al.* [1981, pp. 71-91].

are logically valid. Another consequence of this treatment, combined with clauses (iv) and (v), is that the formula

$$Fa \rightarrow (\exists x)Fx$$

is invalid: consider a world w such that no object in d(w) is F, but a, which does not exist at w, is assigned to the extension of F at w. So the usual rule of existential generalization fails in our system; rules for the quantifiers have to be the rules of free logic (for the empty domain).[7] Note that if we had decided to build the Falsehood Principle into our logic, changing clause (iv) accordingly, the above formula would in general still be invalid; but its instances with *atomic* F, of course, would be valid.

The next step in the development of the semantics is to give evaluation clauses for the logical operators, and here it pays to be rigorous about how the new operators, the objectual quantifiers, are to be handled. Let us step back from modal logic for a moment and consider evaluation clauses for the quantifiers in ordinary predicate calculus. Suppose M is a model for ordinary predicate calculus with domain D. Under what circumstances is an existentially quantified sentence, say '(∃x)(Fx & Gx)', true in M? One standard answer is that there must be some object a in D such that 'Fx & Gx' is *true of it*. Those familiar with this approach, however, will be aware of the complications involved in basing a fully general theory on the notion of 'true of' (those unfamiliar with the approach will be spared the details here). So, instead, we will develop the idea that '(∃x)(Fx & Gx)' is true in M iff there is some object a in D such that the sentence obtained from '(∃x)(Fx & Gx)' by deleting the string of symbols '(∃x)' and substituting some name of a for each free occurrence of the variable x, is itself true in D. More abruptly, if we write '$\underline{a}$' for a name of a and

[7] For a comprehensive account of free logics, see Schock [1968]. To obtain a free logic for the empty domain, it suffices to alter the quantifier rules in Hodges [1977] thus:

(∀): If a formula of the form (∀v)Av occurs on a branch B, any formula of the form E(t) → A[t/v] may be appended to any branch B′ extending B.
(∃): If a formula of the form (∃v)Av occurs on a branch B, any formula of the form E(t) & A[t/v] may be appended to any branch B′ extending B, provided t does not already occur in some formula on B′.

In the context of free logic, the objections in Scott *et al.* [1981] to Hodges' method of allowing for the empty domain no longer apply, since the special restriction Hodges needs on the ∀-rule, that t be a name already in some formula on B′, is not required. The reader may refer to footnote 19 of Chapter 4 for an adequate set of rules for quantified S5.

'A[$\underline{a}$/x]' for the result of substituting the name $\underline{a}$ for each free occurrence of the variable 'x' in the expression A, we can say

'(∃x)(Fx & Gx)' is true in M iff for some object a in D,
'Fx & Gx[$\underline{a}$/x]' is true in M.[8]

Since the expression 'Fx & Gx[$\underline{a}$/x]' stands for the sentence 'F$\underline{a}$ & G$\underline{a}$', we obtain the plausible clause that '(∃x)(Fx & Gx)' is true in M iff for some a in D, 'F$\underline{a}$ & G$\underline{a}$' is true in M. Thus the truth of a certain existentially quantified sentence consists in its having some true singular instance. For example, if Smith, Jones, and Robinson are all the people in the room, then we can say that 'Someone is a spy' is true in the room iff for some a in the room, '$\underline{a}$ is a spy' is true in the room, which in this example reduces to the condition that at least one of 'Smith is a spy', 'Jones is a spy', and 'Robinson is a spy' is true in the room.

However, there is a minor complication to deal with. Consider the sentence

(a) Some planet in the Milky Way is larger than the Sun.

(a) is a sentence of English, but whether or not it is true or false cannot be settled by whether or not it has a true *English* instance of the form 't is a planet in the Milky Way & t is larger than the Sun'. It may well be that there are planets larger than the Sun in other solar systems, but that none of these have names in English, while none of the planets which do have English names is larger than the Sun. The example shows that the suggested approach to the existential quantifier works only if we are evaluating sentences of a language which has a name for every object within the range of the existentially quantified variable. So we shall just assume that we *are* always working with such a language. A model M for a set of sentences X will therefore be understood, from now on, to be a model for the language L of the sentences in X, and this language will contain, in addition to the symbols which occur in the sentences in X, an individual constant $\underline{a}$ for each object a in the domain of the model; so there is really no such thing as *the* language of the sentences in X, for the language changes from model to model acording to the number and identity of the objects in the domain of

[8] In this clause, the first occurrence of the letter 'a' is as a metalanguage variable ranging over objects, but we also wish to use this letter in object language sentences as an individual constant. In such occurrences, it is underlined. Here I follow the convention of van Dalen [1980, p. 68]. When 'a' is being used in both object and metalanguage as a name of a single object, its object-language occurrences are not underlined.

the model. With this explanation, we can now generalize our example to give the following clauses (∃) and (∀) for ordinary predicate calculus:

(∃) A sentence of the form (∃v)Av is true in a model M with domain D iff for some object a in D, Av[$\underline{a}$/v] is true in M.

(∀) A sentence of the form (∀v)Av is true in a model M with domain D iff for every a in D, Av[$\underline{a}$/v] is true in M.

With one more piece of notation, we can proceed with the evaluation clauses for quantified modal logic. Returning to the definition of S5 model just given, let us embody the specification described in clause (iv), which gives the extensions of predicates and relation symbols at worlds, in a two-place function 'Ext'. Thus to say that Ext(H, w) = {⟨a, a⟩, ⟨b, b⟩} is to say that the extension of the two-place relation symbol 'H' at world w are the two pairs of objects ⟨a, a⟩ and ⟨b, b⟩; that is, just the atomic sentences 'Haa' and 'Hbb' are true at w. And let us embody the assignment of a referent to each constant in the language of the model, as described in clause (v), in a function 'Ref'; so "Ref('a') = a" says that the referent of 'a' is a. The evaluation clauses are then as follows:

(vii) an atomic sentence of the form $Ft_1 \ldots t_n$ is true at a world w iff $\langle Ref(t_1) \ldots Ref(t_n) \rangle$ is a member of Ext(F, w);
(viii) an identity sentence t = t′ is true at a world w iff Ref(t) = Ref(t′);
(ix) E(t) is true at w iff Ref(t) is in d(w);
(x) −A is true at w iff A is not true at w;
(xi) A & B is true at w iff A is true at w and B is true at w;
(xii) ◇A is true at w iff for some u in W, A is true at u;
(xiii) □A is true at w iff for all u in W, A is true at u;
(xiv) (∃v)Av is true at w iff for some a in d(w), A[$\underline{a}$/v] is true at w;
(xv) (∀v)Av is true at w iff for all a in d(w), A[$\underline{a}$/v] is true at w.

Two points about these evaluation clauses should be noted, the first concerning the treatment of objectual quantifiers in (xiv) and (xv), the second concerning the treatment of '□' in (xiii). First, the reader should inspect clauses (xiv) and (xv) to see that they embody our idea that the truth of a quantified sentence at a world should turn only on how things are with the objects which exist at that world (unless the sentence contains names of things which do not exist there). Instead of 'exist at that world', we might have used the locution 'are actual at that world', involving a relativized notion of actuality. Because of this terminology, the treatment of the quantifiers in our two clauses is sometimes called the *actualist* interpretation. One justification for the actualist treatment is that it straightforwardly accommodates non-modal quantified sentences; the assertion that everything is made of matter is true *simpliciter* iff all actual, i.e. existent, things, are made of matter, so in this non-modal use the universal quantifier ranges over only the existents of the actual world. Certainly, we do not want to say that that assertion is false because things which might have existed, but do not, are not actually made of matter. So outside the scope of a modal operator, objectual quantifiers are evidently actualist, and (xiv) and (xv) generalize this. Of course, it is easy to add *extra* objectual quantifiers to modal language, the so-called 'possibilist' quantifiers 'Π' and 'Σ', where '(Πv)Av' is true at a world w iff for every a in D (not just d(w)), A[a̲/v] is true at w, but we will in fact have no use for such quantifiers.[9]

The second point to notice is that, according to (xiii), a sentence such as '□Fa' is true at any world iff 'Fa' is true at every world. This is sometimes known as *strong* necessity, in contrast to an interpretation of '□' on which the truth of '□Fa' at w would require only the truth of 'Fa' at all worlds where a exists, an interpretation known as *weak* necessity. One good reason for choosing the strong rather than the weak interpretation of necessity is that use of the latter deprives us of any natural way of regimenting certain modal propositions. The simplest example is from theology, where certain writers have wanted to say that God's existence is non-contingent, or that he exists of necessity, i.e. in every possible world. With strong necessity, this proposition is expressed by '□(God exists)', but when '□' is read weakly, that formula only requires that God exist at every world where He exists, a requirement which not even God can avoid meeting. This is an example of an *expressive weakness* in modal language with weak '□' relative to modal language with strong '□', and many other examples can be given. So (xiii) is to be preferred.[10]

Let us end by illustrating the semantics, first with counterexamples to the formulae (16) and (17) from the previous section of this chapter:

(16) $\Diamond(\exists x)Fx \rightarrow (\exists x)\Diamond Fx$

[9] See Fine [1981b, pp. 192–93] and §5 of the Appendix to this book for further discussion of possibilist quantifiers.

[10] The terminology is from Kripke [1971, p. 137]. The S5 system in Davies [1978] employs weak necessity. See Hazen [1978] for further examples of expressive weakness.

and

(17) $\Box(\forall x)Fx \rightarrow (\forall x)\Box Fx$.

To defeat (16), let W = {u, v}, D = {a, b}, d(u) = {a}, d(v) = D, and suppose that the extension of F is empty (= ϕ) at u and is {b} at v. The picture is:

u .	v .
{a}	{a, b}
Ext(F) = ϕ	Ext(F) = {b}

Choose u for w*. Then '$(\exists x)\Diamond Fx$' is false at w* by clause (iv), since there is no a in d(u) such that '$\Diamond F\underline{a}$' is true at u; in particular, 'Fa' is false at both u and v; on the other hand, '$\Diamond(\exists x)Fx$' *is* true at u, since '$(\exists x)Fx$' is true at v (because 'Fb' is true at v). So (16) is false in this model.

To obtain a counterexample to (17), let W = {u, v}, D = {a, b}, d(u) = D, d(v) = {b}, and for each w, let the extension of F be d(w). The picture is

u .	v .
{a, b}	{b}
Ext(F) = {a, b}	Ext(F) = {b}

Choosing u for w*, we have '$\Box(\forall x)Fx$' true at w* since '$(\forall x)Fx$' is true at u and at v, but '$(\forall x)\Box Fx$' is false at w* because '$\Box Fa$' is false at v in virtue of a's not being in Ext(F, v).

These examples involve rather simple formulae with just one modal operator and one quantifier. To impart a proper 'feel' for the semantics it is useful to consider more complicated arrangements of operators and quantifiers. For instance, we have seen that '$\Box(\forall x)Fx$' and '$(\forall x)\Box Fx$' are not equivalent, but how does the formula '$\Box(\forall x)\Box Fx$' stand to these two? It should be obvious that this formula implies both of the others, so let us consider the converses,

(20) $\Box(\forall x)Fx \rightarrow \Box(\forall x)\Box Fx$

and

(21) $(\forall x)\Box Fx \rightarrow \Box(\forall x)\Box Fx$.

The counterexample immediately above to (17) is already a counterexample to (20), since, in that model, '$\Box(\forall x)Fx$' is true at w* while '$(\forall x)\Box Fx$' is false there, and if '$(\forall x)\Box Fx$' is false at some world, then

'$\Box(\forall x)\Box Fx$' is false at every world. A more interesting question is whether we can have (17) true but (20) still false in a model. The answer is in the affirmative. Let W = {u, v}, D = {a, b, c}, d(u) = {a, b}, d(v) = D, and for each w, Ext(F, w) = d(w).

u .	v .
{a, b}	{a, b, c}
Ext(F) = {a, b}	Ext(F) = {a, b, c}

Here '$\Box(\forall x)Fx$' is true at w* (= u) since '$(\forall x)Fx$' is true at u and v; also, '$(\forall x)\Box Fx$' is true at w* since a and b are in the extension of 'F' at both worlds, so (17) is true in this model, as is the antecedent of (20). But the consequent of (20), '$\Box(\forall x)\Box Fx$', is false at w*, since '$(\forall x)\Box Fx$' is false at v, because '$\Box Fc$' is false at v, since 'Fc' is false at u. Note that in the course of this argument (21) was also refuted.

Consider next the formula

(22) $\Box(\forall x)(\forall y)Fxy \rightarrow \Box(\forall x)\Box(\forall y)Fxy$

and the model where W = {u, v}, D = {a, b}, d(u) = D, d(v) = {a}, Ext(F, u) = {⟨a, a⟩, ⟨b, b⟩, ⟨a, b⟩, ⟨b, a⟩} and Ext(F, v) = {⟨a, a⟩}. Then '$(\forall x)(\forall y)Fxy$' is true at w* (= u) and at v, so '$\Box(\forall x)(\forall y)Fxy$' is true at w*. Let us now evaluate the consequent of (22) step by step. If '$\Box(\forall x)\Box(\forall y)Fxy$' is true at w*, then '$(\forall x)\Box(\forall y)Fxy$' must be true at u and at v; and if this latter formula is true at u, then '$\Box(\forall y)Fay$' and '$\Box(\forall y)Fby$' must both be true at u (by clause (xiii)). But '$\Box(\forall y)Fby$' is false at u since '$(\forall y)Fby$' is false at v, in turn because 'Fba' is false at v. It follows that the consequent of (22) is false at w*, and since we already saw that its antecedent is true there, (22) is refuted by our model. The reader may like to confirm his understanding of this reasoning by constructing a counterexample to:

(23) $\Box(\forall x)\Box(\forall y)Fxy \rightarrow \Box(\forall x)\Box(\forall y)\Box Fxy$.

Lastly, consider the formula

(24) $\Box(\forall x)\Box E(x)$

which in operator discourse reads 'necessarily everything necessarily exists'. The import of this claim is easier to grasp in possible worlds discourse, where we may render it 'in any world, anything in that world exists at every world'. This formula therefore has the effect of ruling out the kind of situation we exploited to construct the various models above, where an element in the domain of one world failed to appear in the domain of another: (24) says that any object appearing in the domain of one world also appears in the domain of every other. It follows from this that in any model M in which (24) is true, the domains d(w) of all the worlds w in W must be the same. In fact, since each object in D is assigned to the domain of at least one world, it follows that the domain of every world is D itself, the set of all possible objects. One variant of the system we have presented adds to the definition of model the condition that the domain of every world is D; we might call this variant 'S5 with unchanging domains' or, for short, 'S5B'. Here 'B' is for 'Barcan', since, as the reader can check, (24) is equivalent to the conjunction of any Barcan formula and its converse. But we will regard the 'unchanging domains' condition as an extra stipulation, like the Falsehood Principle (though without a similar philosophical rationale), and what we have just seen is that this condition can be formulated as a single sentence, (24), of our modal language. Of course, (24) is invalid in our system as it stands—it fails in every model we have presented so far—so when it is necessary to contrast our system with S5B, we shall call it 'S5C', where 'C' stands for 'with contingent existents'.

3. *First order tense logic*

Since we have already laid the groundwork in the sentential case, we might at this point proceed with the introduction of first order versions of the systems T, B (here 'B' is for the propositional system B, not for 'Barcan') and S4. However, as we will not want to use any of these systems in what is to come, developing them would involve too great a digression. Instead, we will pursue the not wholly unrelated question of what happens when we replace our modal operators with operators corresponding to the *tenses* of natural language; for it will turn out that model theory for tense logic (the logic of the tense operators) bears a striking resemblance to model theory for modal logic with an accessibility relation, a resemblance with some philosophical significance.

First, we extract tense operators from the tenses of sentences just as we extracted modal operators from their moods. Consider the sentence

(25) Nixon resigned.

A more cumbersome cognate of (25) is

(26) It was the case that Nixon resigns

which suggests that tensed sentences such as (25) may be thought of as contractions of longer sentences which contain a present-tensed sentence as a part, modified by a tense operator, in this instance, 'it was the case that'. This parallels the idea that subjunctive English sentences are contractions of indicative sentences governed by a modal operator. We will symbolize 'it was the case that' by the operator 'P', and we will also have a future-tensed operator 'F' meaning 'it will be the case that'. It is now possible to formulate a definition of a *general* model M for a language L which is just like a first order modal language, except that it has the operators 'P' and 'F' instead of '□' and '◇'.

A general model M for a first-order language L with tense operators has the following seven components:

(i) a non-empty set T of times;
(ii) a non-empty set D of objects;
(iii) a function d such that for each time t in T, d(t) is a subset of D, intuitively, the objects which exist at t;
(iv) for each n-place predicate F of the language L, and each time t in T, a specification of which n-tuples of objects in D (NOT: 'in d(t)') are in the extension of F at t;
(v) for each individual constant in the language L, a specification of which object in D is the referent of the constant;
(vi) a designation of a particular t in T as t*, the present moment;
(vii) for each time t in T, a specification of which other times t′ in T are preceded by t; we write "Prec(t, t′)", which may be read as "t precedes t′", "t is before t′", "t′ is after t", etc.

Some of the features of this model theory for tense logic help to illuminate corresponding features in the S5 semantics. In place of worlds, we have moments of time, and the changes in the values of d as t changes represent things coming into and going out of existence as time passes. By (v), individual constants are temporally rigid: they have the same reference at each time regardless of what exists at that time. It seems clear that proper names in natural language are also temporally rigid: at all times, 'Julius Caesar' denotes the man who happened to be the Roman conqueror of Gaul. And it is quite reasonable to allow the extension of a predicate or relation symbol at a time to include n-tuples of objects which contain members which do not exist at t. For instance, the two-place relation of being a descendant of holds at the present time between the author and his father, although the latter no longer exists; and this relation will continue to hold of these two persons at later times when neither exists.

To complete the semantics, we have to give evaluation clauses for the logical constants of L, which are just those from the modal case except for the replacement of modal operators by tense operators. So the evaluation clauses for the non-modal operators carry over to the tense logical case, with minor terminological changes: for instance, we are evaluating formulae at times, not worlds. But from the formal point of view this is not a genuine difference. To these clauses we have to add two clauses for the tense operators. These should reflect the intuitive meaning of 'F' and 'P', and should also refer to the relation Prec, which fixes the range of a tense operator in any particular evaluation. So we have

(F) FA is true at a time t in T iff there is some time t′ in T such that Prec(t, t′) and A is true at t′;

(P) PA is true at a time t in T iff there is some time t′ in T such that Prec(t′, t) and A is true at t′.

Given these operators, it is easy to introduce more into our language by definition. For instance, we can have an operator meaning 'henceforth', since 'henceforth, A' may be defined as '—F—A'; we can have an operator 'heretofore', since 'heretofore, A' may be defined as '—P—A'; and we can even keep tense analogues of our modal operators '□' and '◇', which in the context of tense logic have the senses 'always' and 'sometimes'; these operators may either be defined—'□A' means 'A & —F—A & —P—A'—or their modal evaluation clauses (□) and (◇) from §3 of Chapter 1 can be transcribed into tense-logical terminology.

A consequence of our retaining the evaluation clauses for the objectual quantifiers '∀' and '∃' is that when a quantified sentence is evaluated at a time, the range of the quantifier is just the domain of things which exist at that time. So validities and invalidities parallel to those we investigated in the previous section also arise in tense logic. To illustrate briefly, consider the formula

(27) $F(\exists x)Cx \rightarrow (\exists x)FCx$.

It is left to the reader to construct a formal counterexample to (27), combining the methods of §2 above and §3 of Chapter 1, but it is much easier to see the invalidity of (27) at the intuitive level than it was to see the invalidity of comparable modal formulae. For the predicate 'C', substitute 'travels to Jupiter'. Then we can suppose that in the future someone does travel to Jupiter, so the antecedent of (27) is true at the present moment. But the consequent may well be false, for there may not be someone who exists at the present moment who travels to Jupiter in the future: perhaps the first expedition from Earth will not leave until the twenty-second century.

Of course, alternative tense-logical clauses to the ones we have given are also possible. The main candidates are motivated by the idea that the present *and* past are 'real' in a way that the future is not, and that this should be reflected in the semantics. Thus it might be objected, first, that the rigidity of a proper name should consist only in its having a certain denotation x at and after the time at which x has come into existence; secondly, that quantifiers should be allowed to range over not just the existents of the time of evaluation, but also the existents of all previous times; and thirdly, that the Falsehood Principle should be imposed on atomic predicates for times before all members of the relevant n-tuple exist (so the author and his father would not be said to stand in the relation 'is a descendant of' at any time before the author existed).

There is nothing especially unworkable in these suggestions from the technical point of view, but there is a danger of confusion in the grounds which may be advanced on their behalf. For instance, it is clear that 'Julius Caesar' only entered linguistic usage when the name was bestowed, and was not used by speakers to refer to Julius Caesar before he existed (this does not mean that one cannot use a name to refer to an object which does not yet exist—consider the use of 'Brasilia' before the city was built—but only that 'Julius Caesar' was not a name of this sort). However, when we ask whom 'Julius Caesar' denotes in present-day English we are asking about its referent in a particular language, present-day English, and it is eternally true of that language that, in it, 'Julius Caesar' denotes Julius Caesar, for the facts about that language are unchanging; in much the same way, it is eternally true that the Battle of Waterloo was fought in 1815. What is sometimes called evolution of, or change *in*, a language, is best regarded in formal semantics as change *of* language. We make the same point about the analogous modal case in footnote 17 to Chapter 4.

The idea that the quantifiers might be read as neutral between the present and past tense is less objectionable, but it would require the addition of an operator 'Now' or a second existence predicate if we wished to get the effect of an assertion that there exists *now* such-and-such a thing; since the system we have defined is expressively equivalent to the suggested revised system, the former is therefore simpler. The third suggestion, of a Falsehood Principle for the future, provides a good illustration of the idea of genuineness we mentioned in connection with the modal Falsehood Principle. Even if there is some intuitive support for the thought that the author and his father do not stand in the relation 'is a descendant of' at times before the author exists, it is less clear that we have any difficulty in the idea that the author's father and the author stand in the relation 'is an ancestor of' at times before the author exists. But a defender of the future-oriented Falsehood Principle could say that 'is an ancestor of' is not itself genuine, but is analysable into some complex expression involving the future tense and predicates for genuine properties or relations; very roughly, it could be said that 'a is an ancestor of b' holds in virtue of there being some time in the future of the time at which a comes into existence at which such-and-such genuine relations hold between things existing at that time. Nevertheless, however defensible this line of argument may be in this case, the same point applies to tense logic as to modal: the claim that some such analysis is always correct is a philosophical view which should not be built into logic. So our conclusion is that the system of tense logic defined in the preceding clauses has the same preferred status as the quantified S5 system.

Finally, just as it was possible to generate different systems of modal logic by imposing different structural constraints on the accessibility relation, so it is possible to generate different systems of tense logic by imposing different structural constraints on the precedence relation: Prec is the accessibility relation of tense logic. Furthermore, Prec's structural properties correspond rather clearly with intuitive or philosophical doctrines about the structure of time. The most intuitively appealing conception of the time sequence is that it is linear, without beginning and end, and continuous (this means that it contains no gaps: contrast the rational numbers, where there are gaps corresponding to the irrationals, such as $\sqrt{2}$). On this view, the time sequence is formally indistinguishable from the real number sequence, and so if we impose the requirement that the Prec relation be formally indistinguishable from the relation 'is (strictly) less than' on the real numbers, we obtain a tense logic for time as thus conceived; here formal indistinguishability means the indistinguishability that arises if we abstract from the difference in *nature* between a time and a real number. However, there are many alternative structural constraints we could impose on Prec, to obtain discrete time, or time with a beginning and/or end, or circular time, and so on; and we can investigate how changing the structural constraints alters the set of formulae validated by the semantics; but we will not pursue these topics here.[11]

4. *Possibility semantics for quantified S5*

We will complete this part of our discussion of semantics for modal logic by sketching how the possibility semantics we gave for sentential S5 in §4 of Chapter 1 may be extended to quantified S5. Again, our aim is to provide a semantics which validates exactly the arguments validated by the orthodox S5 semantics, this time first-order semantics, our new semantics to permit possible worlds to be incomplete. We are again thinking of possibilities as being completely specified by sentences beginning 'it is possible that', so the resulting incompleteness will have two aspects. First, such a sentence will mention only a few out of all possible objects and use only a few of the predicates of the language; and second, even with respect to the objects mentioned in it and the predicates used, many of the latter will be such that the possibility-specification does not determine whether or not they apply to the mentioned objects. So we will think of a possibility as an entity which has assigned to it, first, a domain of objects; and second, for each n-place primitive predicate F belonging to some subset of the predicates of L, a specification of an extension to F at the possibility, and also of a *counter*extension to F at the possibility, the extensions and counterextensions being drawn from the set D of all possible objects. When an n-tuple of objects is in the extension of an n-place primitive predicate F at a possibility X, this means that F is determinately *true* of that n-tuple at X; and if an n-tuple is in the counterextension of F at a possibility X, this means that F is determinately *false* of that n-tuple at X. In general, for each X and F, many n-tuples of objects drawn from D will be neither in the extension or the counterextension of F at X, and this is how incompleteness arises.[12]

Referents will be assigned to constants of L as before, in a once-for-all manner, so when the extension and counterextension specification

[11] The pioneer of tense logic was A. N. Prior; see, for instance, Prior [1967]. A full survey is given in Rescher and Urquart [1971].

[12] In English, we sometimes use the phrase 'the possibility that A', where the expression A is logically complex. On the present treatment of possibilities, the use of 'the' here is improper: for such an A, there are many distinct possibilities that A, since many distinct assignments of truth-values to atomic sentences may yield possibilities at which A is true.

for a set of predicates is given for a particular possibility, certain atomic sentences are immediately made true or false at that possibility; one might even replace the assignment of extensions and counterextensions to predicates by an assignment of truth-values to atomic sentences, with the same effect. This means that the special concept of possibility semantics, refinement, may be defined as it was in the sentential case: a possibility Y refines a possibility X ($Y \geqslant X$) iff every atomic sentence for which X is defined has the same truth value at Y as it has at X. So let us say that a p-model for quantified S5 has the following components:

(i) a set P of possibilities;
(ii) a set D of objects;
(iii) for each X in P, and for some subset of predicates and relation symbols in L excluding only the identity symbol, an assignment of an extension and a counterextension at X to each symbol in the set; the extensions and counterextensions are drawn from D, and need not involve only, nor all, the members of Ext(E, X); the extension of the existence predicate at X fixes the domain of X, which is why we do not need to specify a function d from possibilities to subsets of D;[13]
(iv) for each individual constant in L, an assignment of a referent from D;
(v) a selection of a particular possibility X in P for p*, the actual possibility; again, we stipulate that the actual possibility must be complete, which is to say that for each predicate and relation symbol of degree n in the language, any n-tuple of objects drawn from D is in either the extension or the counterextension (though of course not both) of the symbol at X.

Finally, we say that a p-model is a structure conforming to (i)–(v) in which the refinement relation induced by the assignments of clauses (iii) and (iv) satisfies the Refinability condition; this, it will be recalled, is the condition that for any atomic sentence $Ft_1 \ldots t_n$, if $Ft_1 \ldots t_n$ is undefined at X in P then there are Y and Z in P such that $Y \geqslant X$ and $Z \geqslant X$ and $Ft_1 \ldots t_n$ is true at Y and false at Z; clearly, this requires the n-tuple of objects $\langle Ref(t_1) \ldots Ref(t_n) \rangle$ to be in the extension of F at Y and the counterextension of F at Z.

As with the sentential case, the complexity of possibility semantics

[13] This course might have been followed in the possible worlds semantics in §2, but would have been less natural there, for one is less inclined to think of non-actual worlds as being 'given' in some verbal specification.

manifests itself in the evaluation clauses. Because we can think of first-order possibilities as assignments of truth-values to atomic sentences, we can take over the clauses for negation and conjunction which we gave in §4 of Chapter 1. But we need new clauses for atomic sentences (including existence sentences) and identity sentences, and also for the quantifiers. Here are the details:

(vi) $F(t_1 \ldots t_n)$ is true at X iff $\langle Ref(t_1) \ldots Ref(t_n)\rangle$ is in the extension of F at X, and is false at X iff $\langle Ref(t_1) \ldots Ref(t_n)\rangle$ is in the counterextension of F at X;

(vii) $t = t'$ is true at X iff $Ref(t) = Ref(t')$;

Clause (vii) does not mention possibilities on its right-hand side, so the same atomic identity sentences are true at every possibility, which ones are true depending only on the assignment of clause (iv). This simply repeats a feature of possible worlds semantics, that because individual constants are rigid designators, the truth-values of atomic identity sentences are settled independently of the contents of the possible worlds.

To deal with the existential quantifier, we have to preserve the usual analogy with disjunction.[14] It will be recalled from §4 of Chapter 1 that:

(viii) $A \vee B$ is true at X iff $\forall Y \geqslant X\ \exists Z \geqslant Y$: A is true at Z or B is true at Z.

Now an existentially quantified sentence can be thought of as equivalent to a disjunction of related sentences with names instead of a quantified variable: our example in §2 above, 'Someone is a spy', illustrated this. More generally, we want to say that the truth of an existentially quantified sentence $(\exists v)Av$ at a world or at a possibility is equivalent to the truth of a single disjunction with multiple disjuncts, each of the form $E(t)\ \&\ Av[t/v]$, where there is one such disjunct for each object in the domain of the world or possibility. This means that if infinitely many objects exist at the world or possibility, then the disjunction is of infinite length; but no matter, for we have already included enough names in the language to name each object; note also the appearance of the existence predicate in each disjunct, which ensures that whether or not the quantified sentence is true at the possibility turns only on facts about objects which would exist were

[14] Or at least, this is the simplest course. I am grateful to Lloyd Humberstone for pointing this out to me.

the possibility realized. Thus a simple way of stating the clause for the existential quantifier is in effect to say that a quantified sentence is true at X iff the associated disjunction is true there, at the same time applying (viii) further to analyse what is involved in the disjunction being true at X. This yields the clause:

(ix) $(\exists v)Av$ is true at X iff for any Y refining X there is some Z refining Y and some name t such that E(t) & A(t) is true at Z (here A(t) is of course Av[t/v]).

Furthermore, it is not difficult to verify that if a formula B is a disjunction, even an infinitary one, each of whose disjuncts is of the form E(t) & A(t), then the negation of B is equivalent to a conjunction (infinite if B is infinite) each of whose conjuncts is of the form E(t) → —A(t). We already have a clause for '&', so if '∀' is introduced by the usual definition '—∃—', we can infer that the following is the correct clause for '∀':

(x) $(\forall v)Av$ is true at X iff for every name t, E(t) → A(t) is true at X.

The reader should pause to confirm that he understand the reasoning here, which will involve working out the clause for material implication by unpacking its definition in terms of '—' and '&'.

Lastly, what becomes of the modal operators? Although we can just take over the sentential clause for '◇', it is not trivial that the sentential clause for '□' is still acceptable. The reader who worked through the proof that □A is true at X (in the sentential case) iff A is true at every Y would have discovered that the following principle is required: for any (not necessarily atomic) sentence A, if A is not true (i.e. false or undefined) at X, then X has some refinement at which —A is true. However, it turns out that this principle is correct in the first order case too, so both modal operator clauses carry forward.[15]

With these clauses in view, the equivalence of possibility semantics

[15] Suppose □A, that is, —◇—A, is true at X. Then for each Y refining X, ◇—A is not true at Y. That is, it is not the case at Y that there is a Z such that —A is true at Z. So there is no Z such that —A is true at Z. Hence every possibility U has some refinement V such that A is true at V. But this does not show that A is true at every possibility U. However, if A is not true at some U, then by the principle mentioned in the text, U has some refinement W such that —A is true at W, and we have just shown that W must have a refinement V at which A is true. Since it is easy to prove by induction that a formula true at a possibility is true at every refinement of that possibility, it follows that A and —A are both true at V, a contradiction.

with the possible world semantics for quantified S5 in §2 above may be demonstrated. As in the sentential case, each world model is already a p-model. To show that for each p-model M there is a world model M′ such that every sentence true in M is true in M′, the simplest procedure is again to take some adequate set of postulates and/or inference rules, such as those given by Fine,[16] and to check that they are all correct for any possibility in any model. By this soundness result, if J is a set of sentences with a possibility model, then J is S5 consistent, and so by the completeness of the postulates and rules, J has an S5 possible worlds model.

It is even clearer in the first order case that possible worlds semantics represents a considerable simplification over possibility semantics. The point of our developing the latter system is not to make practical use of it, but simply to establish that it is in principle possible to do without the condition that the entities with respect to which we evaluate the formulae of a modal language must be complete with respect to that language. The theoretical dispensability of the completeness requirement will be of some interest later.

[16] Fine [1978a, p. 131].

3

The *De Re/De Dicto* Distinction and the Problem of Transworld Identity

1. *Two kinds of formula*

IN EVALUATING formulae of modal and tense logic in Chapter 2, we learnt the significance of the *order* in which operators and objectual quantifiers are arranged. We also learnt that in evaluating formulae which contain operators *within* the scope of objectual quantifiers, the identity of the objects in the domains of the various worlds is important. Thus, if '(∀x)□(x is made of matter)' is to be true at a world w, this requires that the objects which exist at w are made of matter at every world; but if '□(∀x)(x is made of matter)' is to be true at w, the *identity* of the objects made of matter at other worlds is irrelevant, so long as, at each world, all the objects which exist there are made of matter there. So if we consider an arbitrary world u, the truth of the first sentence at w requires that the existents of w be made of matter at u, and so to determine the truth-value of that sentence, we have to be able to ascertain how things are at u with the existents of w; while no such cross-reference between worlds with respect to which objects are which is needed to ascertain the truth value of the second sentence at w. The contrast between the two formulae here is a special case of a more general contrast between formulae which are '*de dicto*' and formulae which are '*de re*'. We can define this distinction in terms of the syntactic structure of formulae:

> A formula with modal or tense operators is *de re* iff it contains a modal or tense operator R which has within its scope either (1) an individual constant, or (2) a free variable, or (3) a variable bound by a quantifier not within R's scope. All other formulae with modal or tense operators are *de dicto*.

Hence '$\mathsf{P}(\exists x)Fx$' and '$\Box(\forall x)Fx$' are *de dicto*, while by (1), '$\Diamond Fa$', by (2), '$\Box Fx$', and by (3) '$(\forall x)\Box Fx$' and '$\Diamond(\exists x)(Fx \mathbin{\&} \Diamond Gx)$', are all *de re.*[1]

The difference between *de re* and *de dicto* formulae, as we see from the example about matter, is a difference between formulae which are, and formulae which are not, sensitive to the identities of objects at various worlds. In evaluating a *de dicto* formula A in a model, we eventually reach subformulae of A whose main connectives are modal or tense operators which have within their scope complete sentences which themselves do not contain any individual constants or modal or tense operators. '$\Box(\forall x)Fx$' is already of this form, and in '$\Diamond(\forall x)Fx \rightarrow \Box(\forall x)Gx$' or '$\mathsf{PP}(\exists x)Fx$' we reach such subformulae after one application of the evaluation rule for the main connective of each formula itself. Having reached such a subformula, one then applies the evaluation rule for the governing modal or tense operator, which in turn will require one to evaluate the formula within the scope of that operator at various worlds or times. This last formula is just a quantified sentence or a propositional combination of quantified sentences, and so in evaluating it at a world or time one is concerned to discover just whether some, or all, of the objects which exist at the world or time, satisfy certain conditions. And this last step can be effected without regard to the *identity* of the existents at that world or time. Note how this semantic account of the difference between the *de re* and the *de dicto* motivates our decision to count sentences with individual constants within the scope of modal or tense operators as being *de re.* For the interpretation of such sentences *is* sensitive to the identity of objects at worlds and times. For instance, in evaluating '$\Box$(Jones is made of matter)' at the actual world, we have to evaluate 'Jones is made of matter' at every world, which requires us to be able to identify Jones at each world.

The distinction is especially clear in the temporal case. If 'F(Someone travels to Jupiter)' is true at the present moment, then the sentence 'Someone travels to Jupiter' must be true later, but there are no constraints on the identity of the person whose travelling to Jupiter at a later time makes the quantified sentence true at that time. But if 'Someone will travel to Jupiter' (i.e. 'There is someone such that in the future, he travels to Jupiter') is to be true now, then this requires

[1] This definition of '*de re*' is what Fine calls the 'strict' sense, and is the natural one when individual constants are treated as rigid designators. See Fine [1978a, p. 143].

that at a later time, t, some person who exists *now* travels to Jupiter *then*, i.e., at t. So we see that in evaluating *de re* sentences, we rely on facts about *transworld* or *transtemporal* identity, facts to the effect that a certain individual at one world or time is identical to a certain individual at another.

2. *Quine's view*

Some philosophers, of whom the most influential has been W. V. O. Quine, have argued that *de re* sentences of *modal* language are problematic in a way in which *de dicto* ones are not.[2] Quine's arguments focus mainly on sentences in which an objectual quantifier binds a variable across a modal operator, sentences which are *de re* in virtue of their satisfying clause (3) of the definition of '*de re*'. Put briefly, Quine's view is that (i) the modal operator '□' is best understood as a disguised predicate of sentences, and (ii) on such a reading, *de re* sentences are illegitimate since quantification into expressions enclosed in quotation marks by a quantifier outside those marks does not make sense. Let us explain these two components of his position in turn.

To say that an operator on sentences is a disguised predicate of sentences is to put forward a hypothesis about the 'real' semantic structure of sentences containing the apparent operator. So Quine is saying that such a sentence as

(1) Necessarily, everything is made of matter

is more perspicuously written as

(2) "Everything is made of matter" is necessarily true.

(2) is a subject-predicate sentence of the simplest sort: it contains a name followed by a predicate. In this case the name is a name of an English sentence, since the effect of putting quotation marks around an expression is to produce a name of that expression; while the predicate is the expression 'is necessarily true'. On our own way of formalizing (1), the adverb 'necessarily' is treated as a sentence operator, but there does not seem to be any very great difference between our method and Quine's. Quine's preference for (2) is based mainly on his preference for truth-functional sentence operators in sentential logic, and as we saw in §2 of Chapter 1, '□' is not truth-functional; but let us not pursue the rationale for this preference here.

[2] See Quine [1961, paper viii] and [1966, paper 13].

(1) is a *de dicto* sentence or, more strictly, receives a *de dicto* formalization, on our approach. But if instead we take a *de re* sentence and apply Quine's interpretation of the modal operator, we obtain something which looks decidedly odd. For instance,

(3) $(\exists x)\Box(x \text{ is made of matter})$

becomes

(4) $(\exists x)$("x is made of matter" is necessarily true).

To see why Quine thinks that (4) is nonsense, consider the following assertion concerning the English word which names the capital city of France, that it contains five letters:

(5) "Paris" contains five letters.

Again, to form a name of the word, we surround it with quotation marks, and then we can make a predication of the *word*. If the quotation marks were deleted from (5), we would still have a subject-predicate sentence, but it would say that that particular French city contains five letters, whatever that might mean. Now suppose that we wish to generalize existentially from (5) to assert that something (some word) contains five letters. Clearly, the correct way to write the result of applying existential generalization is

(6) $(\exists x)$(x contains five letters).

The name in (5) is replaced by the variable 'x', and the name in (5) is not the word 'Paris' but rather the expression ' "Paris" '; possible confusion here arises from the fact that in English a name of an expression contains a display of the expression itself. Suppose, then, that we had fallen prey to confusion, and instead of (6) had written

(7) $(\exists x)$("x" contains five letters).

What does (7) say? It does not say that something contains five letters, since that is the import of (6). The best we can do for (7) is to say that it is composed of a redundant string of symbols, '$(\exists x)$', followed (in redundant parentheses) by a sentence of English,

(8) "x" contains five letters

which falsely asserts of the third last letter of the English alphabet that it contains five letters. The moral is that by surrounding the variable 'x' with quotation marks we form a name of a letter, and even though the letter 'x' is itself displayed *in* its name, it cannot be bound by a quantifier situated outside the quotation marks. However, if we return to (4), we see that such impossible variable-binding is precisely what is there attempted. In fact, construing (4) as we construed (7), we have to say that (4) consists in a redundant string of symbols '(∃x)' followed in redundant parentheses by a subject-predicate sentence of English,

(9) "x is made of matter" is necessarily true.

(9) is presumably false, for the reason that 'is necessarily true' is true only of meaningful sentences, and 'x is made of matter' is not a meaningful sentence. So, by his own lights, it is hardly surprising that Quine rejects all *de re* constructions.

But these difficulties for the *de re* arise from the supposition that we ought to treat '□' as a disguised metalinguistic predicate, a peculiar prescription in view of the fact that there are formally analogous operators which should not be so treated. The sentence

(10) Everything is always made of matter

makes good sense: it is true at a time t iff everything existing at t is made of matter at all times (which most objects are not, since there are times at which they do not exist). Since (10) makes sense, it cannot be rewritten as

(11) (∀x)("x is made of matter" is always true)

since (11) is just as bad as (4) and (7). So the appearance that 'Always' is an operator must be taken at face value, even although it is not truth-functional (for reasons analagous to those underlying the non-truth-functionality of '□'). Why, then, should we baulk at treating '□' as an operator?

Quine has a reason for distinguishing modal operators from tense operators. As soon as such operators are admitted, *de re* sentences can be formed and, as we have seen, evaluation of a *de re* sentence in tense logic presupposes facts about transtemporal identity amongst individuals, while evaluation of *de re* modal sentences presupposes facts about transworld identity amongst individuals. Quine's view is that there is such a relation as transtemporal identity, that is, there are real features of things in virtue of which transtemporal identity obtains or fails to obtain across time between individuals; but the same cannot be said for transworld identity:

. . . our cross-moment identification of bodies turned on continuity of displacement, distortion and chemical change. These considerations cannot be extended across worlds, because you can change anything to anything by easy stages through some connecting series of possible worlds.[3]

These remarks embody the crux of Quine's case against *de re* modality, but this case is apparently not a very strong one, at least in so far as it attempts to *favour* cross-moment identification of bodies. For it is equally true of the temporal case that you can change, if not anything to anything as time passes, at least certain things to startlingly different things, and the philosophical difficulties which arise in virtue of this phenomenon appear to be precisely parallel to those which arise in the modal case. However, we will not go into these matters in detail until Chapter 7; for the moment, let us regard the quoted passage as a *challenge*, a challenge to give an account of transworld identity at least as good as an account of transtemporal identity which appeals to 'continuity of displacement, distortion and chemical change' as criteria for the holding and failing to hold of this relation. What are the modal analogues of these conditions? This is the central question which will be addressed by the later chapters of this book.

For the remainder of this chapter, we will consider what options are available to the philosopher who doubts that an acceptable account of transworld identity is possible, and who therefore doubts the legitimacy of *de re* modal sentences. We shall distinguish three different positions which are motivated by such scepticism about transworld identity:

(I) The first position, Quine's, is one on which *de re* sentences are rejected outright as meaningless. But less nihilistic reactions are possible.

(II) Since the problem is alleged to arise because of a supposed opaqueness in the concept of transworld identity, we could save *de re* modality if we could recast the semantics of quantified S5 so that evaluation of *de re* sentences does not involve a transworld identity relation. Such a recasting, the basic idea of which is due to David Lewis,[4] will be described in §4 of this chapter.

(III) The third position is one on which every *de re* sentence is provided with a *de dicto* equivalent. This may be effected by

[3] Quine [1976, p. 861]. [4] Lewis [1968].

either of two procedures: we can impose restrictions on the class of admissible S5 models, by adding some further clauses to those of §2 of Chapter 2, such that for each *de re* sentence σ there is a matching *de dicto* sentence σ' such that σ' has the same truth-value as σ in each model in the restricted class; or else we can formulate certain principles in modal language from which we can prove that each *de re* sentence has a *de dicto* equivalent. Intuitively, the modal principles would be true in exactly the models counted as admissible by the extra model-theoretic clauses, so they would 'select' this class. One can therefore think of such principles as stating something about the abstract structure of modal reality, that feature of its structure which permits *de re* sentences to have determinate truth-conditions without any presupposition that there are determinate facts about transworld identities and non-identities; the idea is that the truth-conditions of any *de re* sentence would be given by any of the *de dicto* sentences equivalent to that *de re* sentence relative to the principles. We investigate this position next.

3. *Eliminating* de re *modality*

The third position described above is easier to understand in the light of an analogy. Suppose somebody denies that S5C (S5 with contingent existents) is the correct system for broadly logical possibility and necessity, and that his grounds for this denial consist first in the metaphysical view that all actual *and* all merely possible objects are composed out of a given quantity of basic 'stuff', a quantity which is the same in every world; and second, that sentences with quantifiers ranging over ordinary objects made out of portions of this stuff should be reconstrued (for philosophical purposes) as sentences with quantifiers ranging over portions of the stuff. For instance, 'some person is wise' would become 'some portion of basic stuff realizes personhood and wisdom'. Foregoing discussion of the merits of such an analysis, we can see that this view requires the system S5B, in which every world has the same domain of objects, the set of all possible objects; this reflects the idea that the same portions of basic stuff exist in every world, and make up all possible portions of basic stuff; and, according to the analysis, the objectual quantifiers range over these portions. Then this view can be embodied either in an extra clause in the definition of model for quantified S5, a clause to the effect that for each world w in W, d(w) = D, or in the stipulation that the logically valid formulae are those which in the original system S5C are logical consequences of (24) of Chapter 2, '$\Box(\forall x)\Box(E(x))$'. So (24) is an instance of a principle about the abstract structure of modal reality.

Let us now return to the problem of formulating conditions under which every *de re* sentence has some *de dicto* equivalent. That is, we are looking for some set of sentences Σ such that for any *de re* sentence σ there is some *de dicto* sentence σ' such that in S5C we have:

$$\Sigma \models (\sigma \leftrightarrow \sigma').$$

Let us say that such a Σ *permits sentence elimination.* Alternatively, we want some model-theoretic conditions such that for each *de re* sentence σ there is some *de dicto* sentence σ' such that in any model satisfying the conditions, σ and σ' have the same truth-value (here 'model' means 'model for quantified S5', as defined in §2 of Chapter 2). Presumably, one who seeks such a set of conditions, or such a set of sentences Σ, in order to provide *de re* sentences with *de dicto* equivalents, must prefer that the sentences in Σ be themselves *de dicto*, since he regards *de re* sentences as problematic. However, Fine has proved that, in S5C, there is no set of *de dicto* sentences which permits sentence elimination, as defined above. Thus sentence elimination by a *de dicto* set Σ demands that the underlying logic be other than S5C. In fact, Fine has shown that the underlying logic must be a system he calls S5BF, where 'B' is for 'Barcan' (constant domains) and 'F' is for 'flat'. A model is said to be flat iff in each world w in W the individuals all have the same non-modal properties.[5] In effect, this shows that the goal of sentence elimination is unachievable, since S5BF is not a system which it would be reasonable to regard as embodying the logic of our operators for broadly logical possibility and necessity (though this might be disputed by a defender of the view described above which gives a rationale for S5B).

The need for S5BF arises from the requirement that the sentences in Σ all be *de dicto.* Fine suggests that this requirement need not be imposed. Then if we found a set Σ which permits sentence elimination and which contains *de re* sentences, these sentences could be regarded as 'strictly speaking, meaningless. They are merely stipulated to hold in order that the other *de re* sentences may be interpreted by means of

[5] These results are from Fine [1978b]. See Theorems 27 and 28, pp. 299–301.

their *de dicto* equivalents'.[6] It is not obvious that this position is actually tenable but, even if we admit it, the sets of sentences Σ which permit sentence elimination in S5C, our preferred modal logic, are highly unattractive. Fine gives one example of such a Σ, in which there are three sentences, all of them *de re.* We give below the corresponding restrictive conditions on the class of admissible S5C models, since the model-theoretic formulations are easier to grasp (compare the rendering of '$\Box(\forall x)\Box E(x)$' as 'for all w in W, d(w) = D'). In this example, the conditions permit us to eliminate not just *de re* sentences, but also *de re* formulae with free variables. The three conditions are:

(N) For each world w, there are infinitely many objects in D which do not exist in w.

(P) The extension of an atomic predicate or relation symbol at w is drawn only from d(w).

(H′) The model is homogeneous: for any two n-tuples of distinct possible objects drawn from D, if an arbitrarily complex non-modal formula with exactly n free variables is true of one n-tuple at every world, then it is true of the other at every world. Thus, for instance, the necessary properties of any two objects must be the same.[7]

Although this is only one example of a Σ which permits sentence elimination or, rather, an example of the corresponding model-theoretic formulations, it appears to be not uncharacteristic. However, it is highly unsatisfactory that merely in order to legitimize *de re* modality one has to embrace such a curious metaphysical thesis as (N); indeed, it is not at all obvious what independent considerations one might bring to bear to decide (N) one way or the other. And (H′) is even less attractive. Consider the formula

(12) (E(x) → x is not a musical performance).

It might reasonably be held that no human could have been a musical performance and hence that (12) is true of every human at every world. But (12) is false of every musical performance at the actual world and, in general, is true of a musical performance at a world iff that performance does not take place (does not exist) at that world; so musical performances and humans seem to differ as to whether or not (12) expresses a necessary property of them. However, (H′) forbids such differences; and we can see that (12) is indisputably *not* a necessary

[6] Fine [1978b, pp. 277–8].

[7] Fine [1978b, pp. 286–7].

property of musical performances. Thus one who accepts (H′) has to say that for *every* object x, there is a world at which (12) is false of x, that is, a world at which x exists and *is* a musical performance. In particular, each actual human is a musical performance at some world at which he exists. This is a *reductio ad absurdum* of (H′). In conclusion, then, the third position we distinguished as motivated by scepticism about the coherence of the notion of transworld identity appears to lead to the postulation of theses which are wildly at variance with our intuitive judgements about what is possible and impossible. The effort to preserve *de re* modality by the method of providing *de re* sentences with *de dicto* interpretations yields poor results.

4. *Counterpart theory*

In evaluating a *de re* formula such as '□Fa', the role of transworld identity is to determine, for each world w, which object is *relevant* to the truth or falsity of 'Fa' at w. Our evaluation clauses say that 'Fa' is true at w iff the referent of 'a' is in the extension of 'F' at w, so the relevant object is the referent of 'a', which, of course, is a; thus 'Fa' is true at w iff the object which is identical to a is in the extension of 'F' at w. However, one might abstract this notion of relevance from our particular evaluation method, and experiment with other relations besides identity for fixing which object is relevant to a formula at a given world. So the general scheme is that 'Fa' is true at a world w iff the object relevant to 'Fa' at w is in the extension of 'F' at w (is in Ext(F, w)); and we have the option to consider other ways of spelling out 'relevant to "Fa"' besides 'identical to a'.

This is the line of thinking which motivates counterpart theory, originally proposed by David Lewis.[8] Instead of saying that '□Fa' is true at w iff, at each world u, the thing identical with a (at u), i.e. a, is in Ext(F, u), we say (roughly) that '□Fa' is true at w iff, at each world u, the thing which is the counterpart of a (at u) is in Ext(F, u). Note that on this revised account the qualification 'at u' is no longer redundant; only a can be identical to a at u, but perhaps something other than a can be a's counterpart at u. Of course, this change in terminology is futile unless Quine's objection to transworld identity, that there is no acceptable account of it analogous to the account of transtemporal identity, lapses when we introduce the crossworld relation of counterparthood to play the role originally played by transworld identity in

[8] Lewis [1968].

the evaluation of *de re* sentences. But Lewis has an account of counterparthood which, it seems, does cause the objection to lapse. He writes: 'The counterpart relation is a relation of similarity . . . Your counterparts . . . resemble you more closely than do the other things in their world.'[9] So we can give the following criterion:

(C) For x in d(u) and y in d(v), y is a counterpart of x at v only if nothing in v is more similar to x as it is in u than is y as it is in v.

Note, first, that it is consistent with this criterion that x in d(u) has more than one counterpart in d(v), since two or more objects in v may be equally similar, as they are there, to x as it is in u, although more similar than all the other objects in d(v). Second, note that the criterion states only a *necessary* condition for counterparthood between *existents* at two worlds. Why should the condition not be sufficient? The problem is that on any given resolution of similarity, there will always be at least one thing in a world v at least as similar as is anything else in v to x as x is in u. So if the condition were sufficient, every object would have at least one counterpart in every world. However, it is plausible that in some worlds all of the things which exist are *so* dissimilar from a as it actually is that it is difficult to allow even the most similar of these to count as 'representatives' of a at that world, i.e. as a's counterparts there. So we do not want the similarity condition to be sufficient.

The other feature of (C) to remark is that it concerns only existents at worlds. To see the point of this restriction, consider the sentence '□E(a)'. If a is a contingent existent, this sentence should be false, so we need to be able to construct models at whose actual world it is false. In the orthodox semantics, such a model is one with a world at which a does not exist, and the natural translation of this idea into counterpart-theoretic terminology is that a is a contingent existent in a model iff there is some world in the model at which none of the things which are a's counterparts there exist (this is the interpretation of the more general thought, in the terminology of two paragraphs back, that none of the things relevant to 'E(a)' at that world should exist at it). Since (C) concerns only existents, it is consistent with an object's having a non-existent counterpart at a world, and thus leaves room for whatever stipulation we may wish to make to effect this.

These remarks motivate the following reformulation of the criterion

[9] Lewis [1968, p. 114].

of counterparthood into two parts, which together give us a fuller account of a Lewis-style counterparthood relation. First, we say:

(C1) For any object x in d(u), *if* x has a counterpart in d(v) (i.e. a counterpart which exists at v), then for all y in d(v), y is a counterpart of x at v iff nothing in d(v) is more similar to x as it is in u than is y as it is in v.

To complete the account, we must now deal with the case where x has no counterpart in d(v), i.e. no counterpart which exists at v. It turns out that the simplest stipulation is that an object is its *own* counterpart at a world at which it has no existing counterpart:

(C2) For any object x in d(u) and any *distinct* world v, x has no counterpart in d(v) iff x has exactly one counterpart at v and that counterpart is x itself.

We shall need to make a number of other stipulations about the counterpart relation, but we have already said enough to draw attention to a few points. First, on a technical note, it should be observed that our counterpart relation is a *three*-place relation, the relation 'b is a counterpart of a at w', which we write as 'Cbaw'; and according to (C2), this relation can hold even if b does not exist at w, provided b is a itself. Second, since (C2) concerns only distinct u and v, it still leaves open the question of what are the counterparts at u itself of an object in d(u). Third, note also that (C2) strengthens our earlier statement of what it means in counterpart theory for an object to be a contingent existent; in our first formulation, we said this means that at some world none of the counterparts of that object at the world exist at the world, but by (C2) there is only one such counterpart, the very same object; so contingent existence means that at some world *the* counterpart at that world does not exist at the world.

We should not forget that the main motivation for the introduction of counterpart theory is to avoid the problem of having to give some elucidation of the concept of transworld identity, and it may seem that (C1) and (C2) only partly accomplish this. It is consistent with (C1) that, in counterpart-theoretical model theory, the domains of worlds should be disjoint, that is, if there is some x which belongs both to d(u) and to d(v), then u = v. We will in fact impose this condition on models, and we can then deal with the lacuna in (C2) mentioned above by introducing one further case where an object is its own counterpart, that is, at the world where that object itself exists: the disjointness condition guarantees that there is only one such world. But it may be objected that in evaluating a sentence such as '◇—E(a)', (C2) will still require an implicit application of transworld identity, even if it is *called* counterparthood, because the truth-value of this sentence turns on whether a's counterpart at some other world is a itself. The reply to this objection is that (C2) does not involve any concept to which Quine's challenge applies: it is *merely* a stipulation. The relation of transworld identity which requires elucidation, according to Quine, is a relation which holds between *existents* at worlds. Similarly, Quine's account of transtemporal identity applies only to existents at times: it is not part of his view that an object a which exists at a time t is identical to an object b at a later time at which b does not exist iff the requisite continuity of displacement, distortion and chemical change holds between a at t and b at t′ for, obviously, if b does not exist at t′, these conditions *cannot* hold. It is merely a technical convenience to be able to speak of an object at a time at which it does not exist, so as to facilitate the evaluation of sentences about it at that time, and the same consideration applies to the modal case.

Let us now turn to the details of the model theory. What we want to do is to give essentially the same semantics for quantified S5 as we gave in §2 of Chapter 2, except that *de re* sentences will be evaluated using a counterpart relation between objects, and the domains of worlds will be disjoint. The constraint which we should impose on the model theory we are about to construct is that it should contain the standard model theory as a special case. In more detail, this means that if one has a counterpart-theoretic model in which each object in D has exactly one counterpart at each world w in W, then by replacing an object's counterpart at a world by that object itself throughout the model, until every instance of the counterpart relation in the model is also an instance of the identity relation, one should obtain a standard S5 model of the type already defined, and the true sentences of the two models should be the same.[10] When this constraint is met, we can truly say that the

[10] Each standard S5 model M can itself be transformed into an elementarily equivalent counterpart-theoretic model M′ by replacing each object a in D with the objects ⟨a, w⟩ for each w in W. Then every existent a at a world w is replaced by ⟨a, w⟩ so that no object exists at more than one world. The extension of the counterpart relation between two worlds u and v is fixed for the existents of u and v by the sitpulation that ⟨x, u⟩ is a counterpart of ⟨y, v⟩ at v iff x = y; and for each world u and each existent ⟨x, u⟩ at u, if there is no object ⟨x, v⟩ in the domain of v, then ⟨x, u⟩ is ⟨x, u⟩'s one and only counterpart at v. This gives a counterpart-theoretic model M′ and the constraint in the text says in effect that any counterpart-theoretic model M″ isomorphic to M′ should go over

counterpart-theoretic approach generalizes the standard approach: we exchange the transworld identity relation of a standard model for a counterpart relation and then allow this counterpart relation to have different formal properties from those of identity, so that, for instance, an object can have more than one counterpart at a world. But if these are all the changes we make, then certain features of the standard model theory should carry forward. In particular, the treatment of quantifiers should be actualist, '□' should stand for strong necessity, and the underlying non-modal first-order logic should be that of free logic for the empty domain.

A counterpart-theoretic model for quantified S5 for the language L (CTS5 L-model, for short) consists in the following seven components:

(i) a non-empty set W of possible worlds;
(ii) a non-empty set D of possible objects;
(iii) a function d which assigns to each world w in W a subset d(w) of D; d is subject to the constraints that every x in D is in some d(w) *and* that if u and v are distinct then d(u) and d(v) are disjoint;
(iv) for each object x in D and for each world w in W, a specification of which objects in D are x's counterparts at w; this specification is subject to the constraint that if x is in d(u), then for all other worlds v, x is a counterpart of x at v iff no y in d(v) is a counterpart of x at v; furthermore, in these circumstances x is the *sole* counterpart of x at v; and, finally, if x is in d(u), then x is the sole counterpart of x at u;
(v) for every n-place predicate F of L and for each world w, a specification of which n-tuples of objects drawn from D are in Ext(F, w);
(vi) for each individual constant in L, an assignment to it of a referent from D (again, we assume that L has a name for every member of D);
(vii) a selection of a particular w from W to be the actual world w*.

There is a convenient set-theoretic way of thinking of relations which we can apply to (iv). Just as a one-place relation may be thought of as a set of objects, the objects to which the property expressed by the one-place symbol applies, so an n-place relation may be thought of as a set of n-tuples of objects in a fixed order, where the objects in each

into M. From the definition of counterpart-theoretic model to follow, it is clear that this constraint is met.

n-tuple, in that order, stand in the relation in question. In these terms, our counterpart relation is really a set of triples of objects, each triple containing members of D in its first two places and a member of W in its third place. Let C stand for this set; then to say that ⟨b, a, w⟩ is a member of C is to say that b is a counterpart of a at w; clearly, all the facts about the counterpart relation in a particular model are fixed when such a set of triples is given. Furthermore, this set C is itself a subset of the set of all possible triples of objects in which the first two members are from D and the third from W; we denote this set by 'DxDxW'. So clause (iv) says that the counterpart relation is a subset of DxDxW which meets certain conditions; we will see later that it is sometimes useful to talk about relations in this way.

It remains to give the evaluation clauses for the operators in counterpart theory. Here there will be a complication with no analogue in the standard semantics, for, as will be recalled, the counterpart relation is only relevant to the evaluation of *de re* sentences, and then only at a particular stage of the evaluation. For example, consider the sentences:

(13) $\Diamond(\exists x)Fx$

(14) $\Diamond Fa$

and

(15) $\Diamond\Diamond Fa$.

The truth-value of (13) at a world w does not depend on the counterpart relation at any stage in its evaluation ((13) is *de dicto*) but only upon whether '$(\exists x)Fx$' is true at some world. The truth-value of (14) depends immediately on the counterpart relation, since (14) is true at w iff for some u, some counterpart of a at u is in Ext(F, u). On the other hand, the truth-value of (15) does not depend *immediately* on the counterpart relation: (15) is true at w iff for some u, '$\Diamond$Fa' is true at u, and *then* the counterpart relation is relevant. So the modal operator '$\Diamond$' is going to require two clauses, according to whether or not the truth-value of the formula of which it is the main connective depends immediately on the counterpart relation. One can think of *de dicto* formulae like (13) as representing the extreme case of not depending immediately on the counterpart relation, since they *never* depend on it. Suppressing obvious clauses for the sentential connectives, our evaluation clauses are:

(viii) an atomic sentence of the form $Ft_1 \ldots t_n$ is true at a world w iff $\langle Ref(t_1) \ldots Ref(t_n)\rangle$ is a member of Ext(F, w);

(ix) an identity sentence $t = t'$ is true at a world w iff $Ref(t) = Ref(t')$;

(x) $(\exists v)Av$ is true at w iff for some a in d(w), $A[\underline{a}/v]$ is true at w;

(xi) $(\forall v)Av$ is true at w iff for every a in d(w), $A[\underline{a}/v]$ is true at w;

(xii) suppose that of all occurrences of constants in A, $t_1 \ldots t_n$ (not necessarily of distinct types) are exactly those not within the scope of any modal operator in A; then $\Diamond A(t_1 \ldots t_n)$ is true at w iff there is some world u such that for each object $Ref(t_i)$ there is some counterpart c_i of $Ref(t_i)$ at u such that $A[\underline{c}_i/t_i]$ is true at u; here $A[\underline{c}_i/t_i]$ is the sentence resulting from A by replacing each constant-occurrence t_i with a name of c_i, one name per counterpart, $1 \leqslant i \leqslant n$; we do not display any other constant-occurrences there may be in A, and we extend Ref from a function on types to a function on tokens in the obvious way;

(xiii) for all A not comprised by (xii), any constant-occurrence in A is within the scope of some modal operator in A; then $\Diamond A$ is true at w iff for some u, A is true at u;

(xiv) for A as in (xii), $\Box A$ is true at w iff for every u and for every counterpart c_i of $Ref(t_i)$ at u, $A[\underline{c}_i/t_i]$ is true at u;

(xv) otherwise, $\Box A$ is true at w iff for every u, A is true at u.

Here is one illustration of the semantics in action, in a counterexample to

(16) $\Diamond(\exists x)Fx \rightarrow (\exists x)\Diamond Fx$.

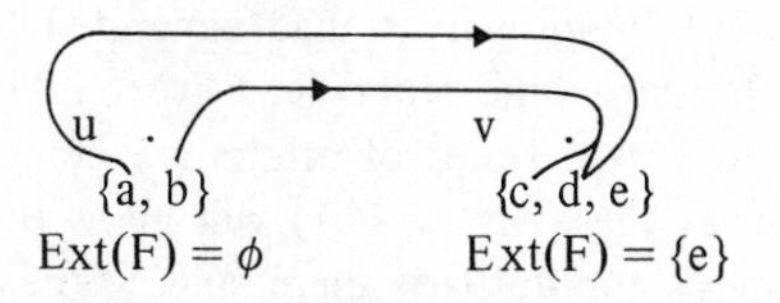

In this model, '$\Diamond(\exists x)Fx$' is true at w* (= u) because '$(\exists x)Fx$' is true at v; but '$(\exists x)\Diamond Fx$' is false at w* since '$\Diamond Fa$' and '$\Diamond Fb$' are both false at w*, in turn because neither a nor b has a counterpart at u or at v which is in the extension of F at u or at v. The counterpart relation is traced by the directed arrows, just as accessibility was in §3 of Chapter 1; so c and d are counterparts of b at v and d is also a counterpart of a at v; this is consistent with the similarity criterion of counterparthood. In standard models, we could also have inserted such lines, tracing the transworld identity relation in the model, but this would have been quite redundant, since the facts about transworld identity are exhibited by the names of the objects in the domains of the different worlds. Nevertheless, such lines are at least implicit, which helps to explain in what sense counterpart-theoretic semantics generalizes standard semantics. As a general term for lines tracing the transworld relation relevant to the evaluation of *de re* formulae, Kaplan speaks of 'transworld heirlines'.[11] In the standard framework, transworld heirlines are one–one since identity is; in the counterpart-theoretic framework, however, the similarity analysis of counterparthood suggests that branching such as is illustrated in our picture should be permitted. Since this appears to be a major departure from the standard framework, there is no reason to expect that the system of validities delivered by the apparatus just set up will be exactly the same system as is obtained on the orthodox semantics. We shall continue to call this latter system 'quantified S5', and the set of formulae valid according to the counterpart-theoretic model theory will be referred to as 'CTS5'.

5. *Objections to counterpart theory*

The version of counterpart theory just outlined improves on Lewis's own version and thus disposes of certain technical objections to his approach.[12] However, some technical and some non-technical objections still remain. A non-technical objection is that it may seem that certain object-language modal sentences which are intuitively true will come out false according to criterion (C1); thus, it seems reasonable to believe that Jones's life could have been very different from the life he has actually led (say, if his parents had emigrated to Australia in his youth) while at the same time others lead lives quite similar to Jones's actual life, i.e. these two states of affairs are compossible. But in a world which realizes these states, (C1) will allow only that the latter individuals be Jones's counterparts there, and so the modal judgement about Jones will be false at the actual world: there is no world in which some counterpart of Jones leads a life very different to Jones's actual life while other individuals lead lives rather like Jones's actual life. Of course, it is possible to reply that in criterion (C1) we are using the notion of similarity in some technical way, in which it does not mean

[11] See Kaplan [1978].

[12] Lewis used a two-place counterpart relation and as a result failed to accommodate contingent existence. See Lewis [1968, p. 119].

just overall similarity in obvious respects. But if the counterpart theorist makes this reply, then it is incumbent upon him to explain exactly what the technical sense of similarity is in which, in our example, the individual in Australia in the imagined world is more similar to Jones as he actually is than any of those other individuals who lead lives in the imagined world which are very similar, in the non-technical sense, to Jones' actual life. At this point, a dilemma arises for the counterpart theorist; for if such an elucidation of the technical sense of similarity cannot be given, then the motivation for counterpart theory, that it employs a crossworld relation less problematic than transworld identity, is undercut; while if such an elucidation can be given, then unless it *entails* that counterparthood is not a one-one equivalence relation, the same elucidation could presumably be applied to transworld identity, which eliminates the motive for developing counterpart theory. We will return to this dilemma in Chapter 7.

A less telling objection to counterpart theory, due to Kripke and Plantinga, is that it misrepresents the content of ordinary modal judgements. On Kripke's view, the counterpart theorist holds that if we say 'Humphrey might have won the election', then

> we are not talking about something that might have happened to *Humphrey* but to someone else, a 'counterpart'. Probably, however, Humphrey could not care less whether someone *else*, no matter how much resembling him, would have been victorious in another possible world.[13]

[13] Kripke [1972, p. 344, footnote 13]. See also Plantinga [1974, pp. 115–16]. Essentially, the trouble with the quoted passage is that its mixture of object-language and meta-language vocabulary makes it sound as if the counterpart theorist allows the truth of 'Humphrey could have won the election' to turn on what happens to other *actual* persons ('someone else') in other worlds. But, of course, this theorist holds that no actual objects exist in any other worlds; he is rather claiming that the truth-value of the remark about Humphrey turns on what happens to certain worldbound individuals in other worlds: Humphrey's counterparts in those worlds. None of these exists in the actual world. This *analysis* of *de re* sentences is not refuted by the mere observation that none of these counterparts is *identical* to Humphrey, for the *point* of the analysis is that it replaces transworld identity with another relation. Similarly, someone might replace continuants and identity through time with instantaneous (timebound) individuals and a transtemporal counterpart relation: it is not *ipso facto* an objectionable consequence of this that the truth of 'Humphrey will win the election' uttered at time t* turns on how things are at a later time t with some instantaneous individual distinct from 'Humphrey-at-t*', for the individual which is relevant at any such time t is still *Humphrey*-at-t. Kripke's attitude to counterpart theory is strange, for in footnote 18 (op. cit.), he goes on to recommend use of the theory to solve a problem similar to Chisholm's Paradox, a recommendation we follow in Chapter 7.

But, as Allen Hazen has forcefully pointed out, this objection, and similar ones due to Plantinga, are quite unfair.[14] According to counterpart-theoretic semantics, the sentence

(17) Humphrey might have won the election

has the truth-condition expressed by

(18) In some world, some counterpart of Humphrey at that world wins the election at that world (in symbols: $(\exists w)(\exists x)$ [Cxhw & Wxw]).

(18) is as good a candidate for being 'about' Humphrey as any sentence of the orthodox semantics; to be sure, there is no mention of counterparts in (17), but there is nothing about worlds in (17) either, although worlds are quantified over in (17)'s orthodox possible worlds truth condition. However, in the quoted passage, it is not the truth-condition which the counterpart theorist *does* ascribe to (17) which Kripke criticizes, but rather one he does not:

(19) Some counterpart of Humphrey could have won the election at some world (in symbols, something like this: $(\exists x)$[Cxhw* & $\Diamond$Wx]).

(19) is not a well-formed sentence of either the modal object language or the counterpart theorist's metalanguage, since it contains both the three-place metalanguage predicate of counterparthood and the modal object language operator '$\Diamond$'; and indeed, (19) does not represent the content of (17). But the counterpart theorist does not claim it does; so the Kripke–Plantinga objections miss the target.

The outstanding technical objection to counterpart-theoretic semantics concerns its handling of the logic of identity. In quantified S5 (the orthodox semantics)

(20) $a = b \rightarrow \Box(a = b)$

is valid. It is tempting to explain why as follows: if the antecedent of (20) is true at w*, then Ref(a) = Ref(b), and since the reference of constants is the same at every world, Ref(a) = Ref(b) at every world and so '$\Box(a = b)$' must also be true at w*. Indeed, (20) would be valid even if we allowed the reference of a constant to change from world to world, provided co-designating constants at one world co-designate at every world. But even the treatment of constants as rigid designators

[14] See Hazen [1979, pp. 319–25].

is in fact not sufficient to guarantee the validity of (20), for if we could make sense of the idea that one object could have been *two*, then at some world which realizes this possibility, 'a' could perhaps attach to one of the two objects and 'b' to the other, so that 'a = b', though true at w*, is false at this world in which the single object at w* is two. But it is clear that on the orthodox semantics with transworld identity, this possibility cannot be represented, and we can see why without reference to the treatment of constants. Note that on the standard semantics we also have the validity of:

(21) $(\forall x)(\forall y)(x = y \rightarrow \Box(x = y))$;

this formula says that if x and y are the same at w*, then they are the same at every world, and again is just a consequence of the fact that one object does not 'become' two or more objects at other worlds. Moreover, this is true for any possible x and any possible y; if we pick an object at one world and the same object at another world, we move along just one transworld heirline. So the following strengthening of (22) is also valid:

(22) $\Box(\forall x)\Box(\forall y)\Box(x = y \rightarrow \Box(x = y))$.

Kripke has argued that these formulae are intuitively valid, so that there would be something *wrong* with a semantics on which they have counterexamples. Consider the example of Hesperus and Phosphorous, otherwise known as Venus. Although it is *a posteriori* that Hesperus *is* identical to Phosphorous (the names were originally associated with different appearances of Venus, and it was a scientific discovery that these were appearances of a single planet) it is surely *a priori* that if these planets are the same, then necessarily they are the same. It is not denied that we can conceive of a world in which 'Hesperus' is a name of one planet and 'Phosphorous' a name of another, but this is not a world in which *Hesperus* and *Phosphorous* are different planets. Since Hesperus, Phosphorous and Venus are all the same planet, a world in which Hesperus is a planet distinct from Phosphorous is a world in which Venus is *two* planets. So the intuitive validity of formulae (20)-(22) turns ultimately on the intuition that *one* thing cannot be *two* (formulae (20)-(22) are versions of a thesis sometimes known as the Necessity of Identity).

As we already remarked, it is consistent with (C1) that a as it is in u has two existent counterparts b and c at v; this would arise if b and c are similar enough to a to be counterparts of a at v, if there is no choosing between them in terms of this degree of similarity, and if they are more similar to a is it is in u than any other object in v. But a model in which this situation is realized is essentially a model of a way in which one thing could be two. The picture is:

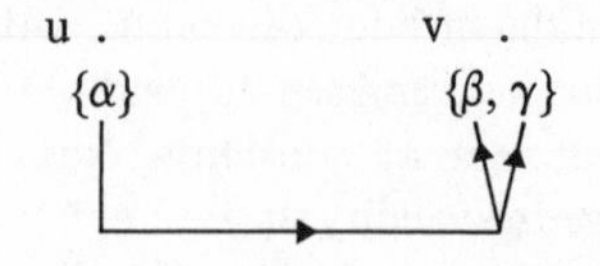

Suppose that 'a' and 'b' are both names of α; then 'a = b' is true at w* (= u). But '□(a = b)' is false at w*, since by clause (xiv), its truth would require that at every world w, any counterpart x of α at w and any counterpart y of α at w are identical, while at v we have both a counterpart x and a distinct counterpart y of α; or putting the same point more precisely in the terminology of (xiv), '□(a = b)' is false at w* because at v, one of the identity sentences containing names of counterparts of Ref('a') at v is false. So we have obtained a counterexample to (20), a counterexample which arises because the counterpart relation need not be one-one. Unsurprisingly, the reader will find that if he allows the counterpart relation to depart from the formal properties of identity in other respects, such as transitivity, further counterexamples to S5 validities involving identity (in the orthodox semantics) can be obtained. There will be more examples later.

It may be suggested that the remedy to this problem is simply to *stipulate*, in the clauses of the counterpart-theoretic model theory, that the counterpart relation is a one-one equivalence relation. But structural stipulations unmotivated by the elucidation of the nature of counterparthood, such an elucidation as (C1), depart from the *raison d'être* of the semantics, which, to repeat, is to provide a model theory which can deal with *de re* sentences without appeal to the allegedly problematic relation of transworld identity. A structural stipulation which goes beyond what is entailed by the elucidation simply imports an unelucidated component into the content of the relation, and Quine's challenge applies again; and there is certainly no case to be made that (C1) by itself entails that counterparthood is a one-one equivalence relation.

A defender of counterpart theory may therefore choose to query the correctness of formulae (20)–(22) themselves. But this has the appearance of a desperate measure, since, as Kripke has pointed out, there is a powerful argument for these formulae.[15] The formula

(23) $\Box(\forall x)\Box(x = x)$

is a validity of quantified S5. But if we combine (23) with Leibniz's Law, according to which, if x and y are the same they have the same properties, we can deduce (20). We can embody Leibniz's Law for a and b in a schema all of whose instances are valid, as follows:

(24) $a = b \rightarrow (\phi(v)[a/v] \leftrightarrow \phi(v)[b/v])$

where '$\phi(v)$' takes arbitrarily complex object-language predicates with one free variable as substitution instances. If we substitute for '$\phi(v)$' any expression which stands for a genuine property of individuals we obtain a valid (not merely true) instance of (24). The property of being necessarily identical to a is a genuine property of individuals, or at least would be said to be so by anyone who regards first-order modal logic as worth doing, and is expressed by any one-place predicate of the form '$\Box(a = v)$'. Substituting in (24) yields

(25) $a = b \rightarrow (\Box(a = a) \leftrightarrow \Box(a = b))$

and (20) follows from (25) *via* (23) (if there is some validity which A implies to be equivalent to B, then A implies B). So a counterpart theorist who proposes to reject (20) must find fault with this argument, which means he must reject (23) or Leibniz's Law. But (23) seems unobjectionable, and Leibniz's Law irresistible.

The upshot of our discussion is this. CTS5 is not the same system as quantified S5, and the difference arises because the counterpart relation, if given a Lewis-style elucidation like (C1), has structural properties inconsistent with those of identity. The counterpart theorist must therefore either find a better elucidation of the counterpart relation, or give reasons why the validities of quantified S5 he rejects should in fact be rejected;[16] and it seems that this second option in turn involves giving reasons why (23) or Leibniz's Law should be rejected. We shall leave matters there for the moment, but in Chapter 7 counterpart theory will again be taken up, and we shall see one rather plausible way in which the counterpart theorist might pursue his case.

[15] See Kripke [1971, pp. 135-41].

[16] A third possibility is that a change may be made in the model theory or evaluation clauses. The reader who wishes to pursue this possibility should consult Hazen [1979] and Fine [1978b, p. 280].

4

Metaphysics for the Semantics

1. *Semantics and explanation*

IN THIS chapter, we begin the investigation of an assortment of philosophical problems which arise in connection with modality. We will be concerned mainly with the philosophical *justification* of a variety of modal theories or theses whose *meaning* will be assumed to be well-understood. But in view of the material of the previous chapters, the first philosophical issues to demand our attention are issues about the semantics of the modal operators themselves. These issues are sufficiently general not to turn on which of the various approaches already presented we adopt, so we restrict our attention here to the orthodox possible worlds semantics for S5, without accessibility.

We have regarded possible worlds semantics as a tool for fixing the powers of the logical operators, for determining which modal logical arguments are valid and which invalid. We saw that there is not always a unique answer to the question 'Valid or invalid?', for there are some arguments, involving iterated modalities, about which we perhaps have no very firm intuitions, which are valid in some systems and invalid in others. But there are also central cases, which any semantics has to get right if it is to be taken seriously; for instance, there is the example already discussed of:

$$\text{(A)} \qquad \frac{\Diamond P \quad \Diamond Q}{\Diamond(P \,\&\, Q)}$$

However, it would be misleading to suggest that the authority of possible worlds semantics derives *merely* from its getting the cases about which we do have firm intuitions right, its dictates about the peripheral cases being a matter of indifference. There are algebraic approaches to questions of validity which will also do that[1] but, in comparison with

[1] See Hughes and Cresswell [1968] Chapter 17. For more advanced use of algebraic techniques, see Goldblatt [1976a, b].

these, there is undeniably a sense in which possible worlds semantics is the 'natural' semantics. But in what, precisely, does this naturalness consist? A tempting reply to this question is that the naturalness arises out of the treatment of the modal operators as quantifiers over possible worlds: there must be a sense in which this treatment is the correct treatment. It is in virtue of this that we can say that the possible worlds semantics does not merely *pronounce that* (A) is invalid, it also *explains why* it is invalid: we understand what is wrong with (A) when we are introduced to the existential quantifier treatment of '◇', which engages our prior understanding of what is wrong with:

$$\text{(B)} \qquad \frac{(\exists w)P(w) \qquad (\exists w)Q(w)}{(\exists w)(P(w)\ \&\ Q(w))}$$

The philosophical problem for this view is to elucidate the sense in which the quantifier treatment is correct, in such a way that the invalidity of (A) is explained by relating it to (B).

The most obvious suggestion about the sense in which the quantifier treatment is right is the suggestion that in translating a sentence such as:

(1) ◇P

by the principles employed in Chapter 1 into the sentence:

(2) $(\exists w)P(w)$

we are translating one sentence into another with the *same meaning* (recall that we read (2) as 'there is some world w such that P holds at w'). Then the relationship between (A) and (B) which permits the invalidity of (B) to explain the invalidity of (A) would just be that the sentences in (A) mean the same as their translations in (B). So the quantifier treatment is right because it maps sentences into synonyms (since this is intended to be a substantial claim, we shall refer to the possible worlds translations of modal sentences simply as their renderings, which is a more neutral term than 'translation' *vis à vis* the question of synonymy). However, the correctness of the quantifier treatment cannot consist just in its preserving meaning: there must be an asymmetric element in this synonymy relationship, otherwise we would not be able to say that the invalidity of (B) *explains* the invaidity of (A). '◇' is the mysterious operator, the one whose logical powers are being investigated, while the existential quantifier is already understood.

To capture this idea of asymmetry, let us say that (2) 'articulates' or 'exhibits' the 'real' meaning of (1). On this view, the significance of the quantifier treatment of modal operators is akin to the significance which philosophers have generally attributed to regimentations of 'problematic' sentences in standard first-order logic. A classic example is that of Russell's Theory of Definite Descriptions. Definite descriptions, phrases of the form 'the F', appear to be terms for referring to objects, much like proper names. For various reasons, Russell wished sharply to distinguish descriptions from genuine names and so proposed an analysis on which such a sentence as:

(3) The girl next door is blond

is said to have the 'real' meaning

(4) There is exactly one girl next door and she is blond.[2]

According to Russell, (3) has the 'surface' structure of a subject-predicate sentence in which the subject term is 'the girl next door', i.e. the sentence has the same structure as 'Marilyn is blond'. If we use the symbol 'ι' to form definite descriptions, reading '$(\iota x)F(x)$' as 'the x which is such that F(x)', or, simply, 'the F', then (3) would be formalized according to its surface structure as:

(5) $B[(\iota x)Gx]$

a sentence with the same form as 'Bm', while (4), of course, becomes:

(6) $(\exists x)(Gx \; \& \; (\forall y)(Gy \rightarrow x = y) \; \& \; Bx)$.

Thus the surface structure of (5) is misleading as to its 'real' meaning, which is exhibited by (6). In particular, (6) shows that (5) does not really contain a subject term (and hence (4) shows the same about (3)) since there are only quantifiers, predicates and connectives in (6); this is the result which Russell wanted. Furthermore, by attributing the logical form of (6) to (5), we can explain the logical powers of the operator 'ι' from which definite descriptions are formed. For instance,

(7) There is a girl next door

appears to follow from (3), intuitively speaking, and if (3) means what (4) means, we have an explanation of why the inference is valid. The suggestion is, then, that possible worlds semantics explains validity and invalidity in the same way; that is, the relationship between (1) and (2) is the same as that between (5) and (6).

[2] See Russell [1918, VI] and Kalish, Montague and Mar [1980, Chapters VI, VIII].

Before looking at some of the consequences of this view, let us pause to ask if there is any relation between (1) and (2) weaker than synonymy which could do the same job. Our minimum requirement is that any candidate R be such that when modal sentences and possible worlds sentences stand in R, then a modal argument and its R-corresponding possible worlds argument are either both valid or both invalid. More precisely, we insist on a relation which meets this condition: if σ is a sentence of modal language and σ' its rendering in possible worlds language according to the quantifier treatment of modal operators, then the hypothesis that σ stands in the candidate relation to σ' should be sufficient to guarantee that σ behaves the same way in a modal argument as σ' does in the rendering of that argument in possible worlds language. Thus, for instance, we are asking whether there is some relation other than synonymy such that if '$\Diamond$P' and '$(\exists w)P(w)$' stand in it, then it follows that the behaviour of '$\Diamond$P' in the argument (A) displayed above is the same as the behaviour of '$(\exists w)P(w)$' in (B). The idea of 'same behaviour' here is still rather intuitive (we will make it more precise later), but the sense of the question is clear enough to see that there is one rather trivial answer to it: we can simply *define* a relation of behavioural equivalence as that relation which two such sentences stand in when they behave in the same way in pairs of corresponding arguments like (A) and (B). Thus, if our possible worlds renderings of modal sentences map sentences onto behavioural equivalents, it is trivially true that a modal argument is valid iff its corresponding possible worlds argument is valid. But it is quite clear that the fact that this relation holds between modal sentences and their possible worlds renderings does not ground the ability of possible worlds semantics to *explain* the validity and invalidity of modal arguments. For it would be an equally substantial question *why* our method of producing possible worlds renderings *via* the quantifier treatment of the modal operators *does* yield behaviourally equivalent sentences; and again, the answer which strongly suggests itself is that a possible worlds rendering of a modal sentence σ is synonymous with σ. If this answer is incorrect, then the semantical systems of the earlier chapters may have done no more than engender an illusion of understanding.[3]

So the view at which we have arrived is that possible worlds semantics explains validity and invalidity because (a) the quantifier treatment of the modal operators produces synonyms, and (b) a possible worlds rendering of a modal sentence exhibits the real meaning of that

[3] The phrase is Quine's. For further discussion, see Scott [1971].

sentence (the synonymy relation has a preferred direction). But this view has a disturbing feature, in that the quantifier treatment is *ontologically radical*: it introduces entities of a certain sort, possible worlds, which are apparently not introduced by modal sentences themselves. At this point, the analogy with the Theory of Descriptions breaks down, for Russell's motivation for that theory was a certain kind of ontological conservatism. Briefly, Russell held that a subject-predicate sentence would be meaningless if its subject term did not succeed in picking out some object. Now, if one holds that sentences with the structure (5),

(5) $B[(\iota x)Gx]$

are genuine subject-predicate sentences, then when faced with such a sentence as 'the present King of France is bald', the Russellian must either say that the sentence is meaningless, which flies in the face of the facts, or that there is such an entity as the present King of France, which also appears to fly in the face of the facts. But one could grasp the second horn of the dilemma and say that there is such an entity as the present King of France, a non-existent entity, and that there are non-existent objects generally. This would be an ontologically radical move. Another way, preferred by Russell since he did not wish to introduce non-existent objects, is to prevent the argument to the dilemma from getting started, by denying that 'the present King of France is bald' is really a subject-predicate sentence.

In our case, we are moving from the ontologically conservative (1), '$\Diamond P$', to the ontologically radical (2), '$(\exists w)P(w)$', although we could say that the explicit ontological commitment in (2) is at least implicit in (1). Furthermore, there does not appear to be any way of avoiding this commitment, if possible worlds semantics *explains* validity and invalidity for modal arguments. (2) says that *there is* a possible world of a certain sort; if this is not literally true while (1) is literally true, then the invalidity of the argument (B) is irrelevant to the question of whether (A) is valid or invalid, since sentences in (B) do not mean what sentences in (A) mean, on the view of explanation of meaning we are presently canvassing. So let us accept the extra ontology apparent in possible worlds discourse; let us agree that well-formed instances of (2) are literally true or literally false, and that there are some literally true instances, since there are some literally true instances of (1). That is to say, we agree that there are possible worlds. We shall say that by this agreement, we are *realists* about possible worlds, since we take them to be real things.

We may distinguish *absolute* realism from *reductive* realism. An absolute realist is one who holds that the notion of a possible world cannot be further analysed; so (2) is as far as we can go in exhibiting the content of (1) in better understood terms. David Lewis is the absolute realist *sans pareil*, but his position includes two extra ingredients which are not essential to absolute realism. First, Lewis holds that each possible world is a thing of the same kind as the actual world, and second, that physicalism is true of the actual world. Hence every possible world is a physical system.[4] But other views are conceivable, on which possible worlds are some kind of *sui generis* abstract object; and an exception might, or might not, be made of the actual world.[5]

A reductive realist is one who holds that possible worlds can be identified with (constructions out of) other entities, themselves held to be less problematic than worlds. Three such positions are that worlds are maximal consistent sets of propositions (or propositions of a certain sort), that they are maximal states of affairs, and that they are maximal possibilities.[6] It follows from this that the absolute/reductive distinction is not the same as another common in the literature, between actualism (according to which only actual things exist) and possibilism. The absolute realist is a possibilist, as is a reductive realist who identifies worlds with certain kinds of possibility, but if all the propositions and sets of propositions which there are actually exist, then the view that worlds are certain sets of propositions or propositions of certain kinds, is an actualist view.

Either variety of realism about possible worlds is of course opposed to anti-realism about them.[7] An anti-realist says that worlds do not

[4] See Lewis [1973, pp. 84–91]. [5] See Davies [1981, p. 200].

[6] That worlds are sets of propositions is argued in Adams [1981]; that they are propositions of a certain sort in Prior and Fine [1976, pp. 116–78]; that they are maximal states of affairs in Plantinga [1974, pp. 44–5]; and that they are maximal possibilities in Humberstone [1981]. An idea which has not been worked out in the literature is that worlds are mental constructs of some sort (McGinn [1981]); this would be a version of reductive realism about incomplete possibilities. See § 2 of Chapter 9.

[7] I am not using the label 'anti-realism' in its contemporary sense to mean a position which denies that the meanings of undecidable sentences are their (*ipso facto* verification-transcendent) truth-conditions. For an explanation of this kind of anti-realism, see Dummett [1975b]. My use is the old-fashioned one, in which an anti-realist about F's is one who denies the existence of F's. That these two notions of anti-realism are not unconnected has been argued recently by Dummett in [1981]. See especially pp. 66–9, where Dummett characterizes a view such as ours (according to which there really are no such things as worlds) as springing 'from the perception of a genuine and important fact . . . that we do not need to invoke the notion of reference, as applied to such terms [for possible

exist, and thus he is an actualist of a more radical kind than any reductive realist. For an anti-realist, any possible worlds sentence which has an existential quantifier over worlds as its main connective must be strictly and literally false; since he will hold the non-existence of worlds to be necessary, this formulation is correct even if a possibilist existential quantifier is admitted (recall that the possibilist quantifier 'Σ' has the clause that '$(\Sigma v)Av$' is true at w iff for some a in D, A[a/v] is true at w, i.e. there is no restriction to those a in d(w)). So the anti-realist cannot take the attitude towards possible worlds semantics outlined above: he cannot say that possible worlds sentences exhibit the real meanings of modal sentences in a peculiarly perspicuous way. The most interesting philosophical question about the semantics of modal logic is whether it is possible to develop an anti-realist view that is consistent with our intuition of naturalness in the quantifier treatment of the modal operators, and which can deal with the thought that the invalidity of (A) is somehow explained by the invalidity of (B). But if some kind of realism were quite satisfactory, this question would be purely hypothetical; to give it some practical urgency, then, let us see whether there are reasons to have qualms about realism.

2. *Realism about worlds*

A very natural consideration in favour of absolute realism about worlds arises from semantic parallels between tense and modal operators. In tensed languages, according to the analysis presented in §3 of Chapter 2, we have operators not obviously of quantificational form which manifest themselves in the surface structure of English as tenses of verbs; we also have explicitly quantificational expressions such as '*some*times' and 'always'; and the tenses are treated as relativizing the semantic values of the expressions on which they operate to entities (times) over which the quantificational expressions also range. In the modal case, we treat the subjunctive mood in surface English analogously, where explicitly quantificational expressions such as 'in all possible circumstances' function in a manner similar to that of 'sometimes' and 'always'. These syntactic parallels might also be extended

worlds], in order to explain how a sentence containing such a term is determined as true or false . . . an understanding of those statements [e.g. possible worlds statements] involves an implicit grasp of their relation to statements of the reductive class' (which in our context is the class of modal statements). I have no quarrel with this semantic conception of what I am calling 'old-fashioned' anti-realism.

to include expressions for places; although nothing corresponds to tense or mood, there are quantificational expressions like 'everywhere' and names of places like names of times (dates). Moreover, in English there is a variety of spatial and temporal indexicals like 'here', 'there', 'then' and 'now', whose reference in a particular utterance is determined by the place or time at which the utterance is made. Someone impressed with the parallel drawn so far may then press it further by suggesting that 'actually' plays a similar indexical role, its reference in an utterance being the world of utterance.[8] Thus there exist the materials for the view that realism about worlds is as well motivated as realism about places and times. Just as we can speak of places and times forming their own manifolds or spaces, so we can say that worlds are the points of a logical space.

There can be no objection to the introduction of such a metaphor, but it supports realism about worlds (granted realism about places and times) only if the similarities upon which the metaphor relies for its appropriateness relate features of logical space to features of space and time manifolds which themselves are inconsistent with anti-realism about places and instants. However, crucial features of places and times which appear to underpin the plausibility of realism about these entities have no parallel in the logical space of possible worlds. For places and times, there is a distinction between the item and its occupier, a material object in the case of a place, and an event in the case of a time. (Strictly, it is regions and intervals which are occupied by objects and events, but places and instants can be 'abstracted' from these; and it is reasonable to hold that possibilities and refinement correspond to regions and abstraction.[9]) It seems crucial to our ability to distinguish places and times from their occupiers that we have the conception of the *same* place, or time, being occupiable by something distinct from its actual occupier; even in the temporal case, someone who denies that a particular token event e which occurs at a time t could have occurred

[8] The treatment of 'actually' as a context-dependent operator is developed by Lewis in [1970] and integrated into a general theory of context-dependence in Kaplan [1977]. For further discussion, see Adams [1974], Davies [1983] and Forbes [1983].

[9] The general method of obtaining points of an n-dimensional manifold from regions of the manifold is called 'extensive abstraction'. For an account, see Tarski [1956, pp. 24–9]. A possibility is a proposition which holds at a number of worlds, thus defining a set of worlds, i.e. a region, like a set of points. As the proposition is refined to include more and more information, the region gets smaller as worlds drop out, until we reach one of Fine's 'world propositions', i.e. a possible world, or a point in logical space.

at another time, will not deny that events could have occupied t other than those which do occupy it. However, this conception is quite inapplicable to logical space; given a world, one cannot distinguish a location and a content contingently located there, no matter which component one identifies with the world itself. How does this difference arise?

It is apparently sufficient for the distinction between location and occupier to be applicable that there be some contingent relational structure amongst the occupiers which either *determines* or is *determined by* the locations of the occupiers (for a relationalist, a type of reductive realist about space, these relations determine the locations of objects, while for an absolutist, the converse is true).[10] To see that relations weaker than determination may not be sufficient, consider the case of colour space.[11] Any colour can be regarded as a combination of red, green, and blue in specific intensities with numerical values x, y, and z respectively. Thus any colour can be given coordinates $\langle x, y, z \rangle$ with respect to the three axes red, green, and blue. Moreover, we can define distance on this three-space in such a way as to reflect real phenomena of colour perception. If we say that a *threshold of distinction* for a given colour is the amount of continuous alteration needed before a human being perceives a change in colour, then the distance between two colours can be identified with the smallest threshold of distinction which can be laid between them. This distance relation is contingent, since humans could have had better or worse powers of sensory discrimination. Nevertheless, we do not seem to be able to abstract a colour space from the colours which fill it, even though in some sense this space has actually unoccupied regions, such as Hume's missing shade of blue. The problem is that the distance relations and the coordinates of the colours are quite independent: there is no inclination at all to think that if the distance relations had been different, a different colour would have had the coordinates actually possessed by, say, the colour of the jerseys of the Tulane Green Wave.

There are no natural relations on possible worlds corresponding to distance relations, and although relations could be introduced, such as the relation 'w is more similar to u than is v' (for a fixed method of resolving respects of similarity), there is nothing contingent about such relations. Hence there is no means by which we might distinguish

[10] See Forbes [1984b] for more about the contrast between the absolutist and the relationist.

[11] See Aleksandrov *et al.* [1983, vol. 3 pp. 151–3].

a possible world from what is true at it: the content of a world is in no sense something which occupies the world. And this means that the appropriateness of the metaphor of logical space does not reside in similarities which motivate equal degrees of realism about places and times, on the one hand, and worlds, on the other. For our ability to separate a place, or a time, from its occupier is crucial to realism about places and times, be it absolute or reductive, as is the applicability of a distance relation to the places and times themselves. The means by which the conceptual separation is effected is by holding the distance relation between the points constant while changing the distance relation between the occupiers (by moving them around, or by deletion with or without replacement). This procedure attributes necessity to the facts about distances between points themselves, which gives them identity criteria and therefore 'objecthood' independent of that of the category of occupiers. But this apparatus is not available for worlds.[12]

However, a reductive realist who is also an actualist will find this limitation of the metaphor of logical space not particularly dismaying, since his realism about worlds is derivative from his realism about the entities from which he holds worlds to be constructed. A more general objection has to be pressed against such a realist, and one who is unpersuaded by the objection we are about to produce will find the anti-realism which avoids it concomitantly unmotivated. The main objection against both sorts of realism about worlds is the nominalist–actualist objection from epistemology. According to this objection, knowledge of properties of objects requires experience of these objects or of their effects, which in turn requires that these objects or effects be within the range of our sensory faculties. But only objects which are both concrete and actual are, or have effects which are, within the range of these faculties. However, the realist holds that the expression '$\Diamond$P' attributes a property, that of P's holding, to an object, a world, which is non-actual according to the absolute or possibilist-reductive realist, and non-concrete according to the actualist reductive realist. Hence realism renders it generally impossible to know whether or not '$\Diamond$P' is true.

The strength of this objection depends upon the plausibility of nominalism and actualism, but the problem of how it is possible that we have knowledge of propositions from certain areas of discourse is

[12] In [1981] McGinn argues that this failure of analogy constitutes a reason to reject the reality of possible worlds. This paper contains an interesting defence of anti-realism about worlds conjoined with realism about modal reality.

a powerful consideration in favour of these positions.[13] Since we cannot conduct a general discussion of the issues here, let it suffice to say that whatever force the epistemological objection has is reflected by a corresponding urgency in the development of an adequate anti-realist view of possible worlds.

3. *Two problems for anti-realism*

The challenge for the anti-realist is to give an *interpretation* of the appealing features of possible worlds semantics which shows how these features can arise even though there are no such things as worlds; he cannot just ignore the semantics, given the intuitions we have about its naturalness, for this is a phenomenon which surely requires explanation. Furthermore, his interpretation must posit *some* semantic relationship between sentences of modal language, which we shall call 'L_m', and their renderings in possible worlds language, which we shall call 'L_w'; for without such a relationship, it must seem positively miraculous that the semantics agrees with our intuitions about validity and invalidity. And we saw earlier that it is hard to think of any candidate for this relationship other than synonymy. It follows that the component of the pro-realist view about the semantics which the anti-realist must attack is the claim that the synonymy relation is asymmetric in the direction which makes the possible worlds sentences stand to modal sentences as Russell's interpretations of the sentences of a language with the operator 'ι' stand to the sentences of that language: the anti-realist has to allow synonymy, but deny that the possible worlds renderings exhibit the real meanings of the modal sentences. In particular, since he is an *anti*-realist about worlds, he has to say that objectual quantifiers, when they range over possible worlds, do not have their literal meaning, the meaning they have in ordinary first-order languages; in turn, then, the sentences of possible worlds language do not mean what they appear literally to mean.

What, then, is the meaning of such sentences? The simplest manoeuvre available to the anti-realist here is to reverse the direction of the asymmetry in the synonymy relationship. Instead of saying that the meaning of a modal sentence is given by its L_w rendering, we can say that the meaning of an L_w sentence is given by its rendering in (reverse translation into) L_m; so in the simplest case we say that (2):

(2) $(\exists w)P(w)$

[13] See Benacerraf [1965] and Field [1980, pp. 1–19].

should be understood as having the meaning (1) has,

(1) $\Diamond P$

or that (2) has a meaning imputed to it by (1). Since there is no literal assertion of the existence of possible worlds in (1), it follows that there is no literal assertion of the existence of worlds in (2), despite appearances. However, there are two outstanding problems for this anti-realist position. The first problem is the problem of validity. If L_w sentences have non-literal meaning, there must be an element of the incidental in the methods of possible worlds semantics for determining validity and invalidity: there has to be a more direct method. Furthermore, whatever this method is, one has to be able to derive from it an explanation of *why* possible worlds semantics is successful. We have said that the invalidity of (B) in some sense explains the invalidity of (A), and in deriving the invalidity of the latter from that of the former, we are assuming, by anti-realist lights, that however the meaning of the sentences in (B) differs from the apparent meaning they have, the meaning which results from interpreting their quantifiers literally, this difference does not change the logic of the quantifiers in these non-literal occurrences. The fundamental account of validity for L_m must justify this assumption; and if it can succeed in doing this, we can account for the intuition that the semantical status of (A) is illuminated by its rendering as (B) simply in terms of the great familiarity of first-order languages.

The second problem which faces the anti-realist doctrine we are investigating is the problem of reverse translation. In proposing that each possible worlds sentence be ascribed the meaning of the modal sentence of which it is a rendering, we are proposing an elimination by paraphrase of the ontology of possible worlds, as opposed to a reductive identification of worlds with other entities. But this elimination is possible only if every meaningful possible worlds sentence is a rendering of some modal sentence, and as L_m and L_w presently stand, this is not so. In fact, much of the difficulty lies in an expressive weakness in L_m, a weakness which the anti-realist must show how to remedy to make his position plausible. We deal with these problems in turn in the next two sections.

4. *Validity: other approaches*

To justify attribution of normal logic to quantifiers when they bind world variables in L_w sentences, we have to show that the valid/invalid classification consequent upon this attribution is in agreement with the classification delivered by criteria which apply directly to L_m sentences, without detour through non-literal renderings of them. So, first, we have to find criteria of this sort; and there are two types of criteria we might hope to develop, criteria from proof theory and criteria from alternative semantics. Let us begin with proof-theoretic criteria.

In developing possible worlds semantics in Chapters 1 and 2, we assumed a fund of intuitions about the correctness or incorrectness of particular arguments. What precisely is the source of these intuitions? One possible account is that competent speakers of English have native intuitions about what *follows* from what in their language, and when presented with the argument schemata of formal logics, intuitions about the particular connectives occurring in these arguments are isolated and activated, so that it is just obvious to such a speaker, at least in the simple cases, whether or not the conclusion of any English instance of the schema would follow from its premisses. According to this view, on which the bedrock 'pretheoretic' intuitions are intuitions about what follows from what, the fundamental method of encapsulating the meaning of a logical constant is to give a rule for when a sentence with that constant as its main connective follows from other sentences, and also a rule which determines what other sentences follow from it. So this view looks to the 'natural deduction' rules governing a connective for the embodiment of the essence of a native speaker's mastery of the connective.[14]

To give a simple illustration, consider the connective '&'. The typical semantic account of validity for propositional languages presupposes that the meaning of '&' is fixed is some semantic way, usually by a truth-table or fundamental truth-value matrix. On the present view, the meaning of '&' should rather be given by an introduction and elimination rule for the connective, thus:

&-Introduction: if A has been proved from premisses X and B has been proved from premisses Y then A & B follows from premisses $X \cup Y$;

&-Elimination: if A & B has been proved from premisses Z, then A follows from Z and B follows from Z.

[14] The main proponent of this view was Gentzen. See Gentzen [1969, Chapter IV] and also Prawitz [1971]. There is a useful brief survey of some recent work in Grandy [1982].

More formally, the rules may be written:

&-I: if $X \vdash A$, $Y \vdash B$, then $X \cup Y \vdash A \& B$.

&-E: if $Z \vdash A \& B$, then $Z \vdash A$ and $Z \vdash B$.

In terms of these rules, we can justify the truth-table for '&' in the light of a certain view of the truth-predicate associated with Quine. According to Quine, although use of the truth-predicate involves 'semantic ascent', so that instead of making a statement about the world we predicate a property of a sentence,

> . . . the truth predicate serves, as it were, to point through the sentence to reality . . . Thus ascent to a linguistic plane of reference is only a momentary retreat from the world, for the utility of the truth predicate is precisely the cancellation of linguistic reference . . . The truth predicate is a device of disquotation.[15]

Suppose we put 'P&Q' for Z in &-E. Then, since $P \& Q \vdash P \& Q$, we infer (by &-E) that $P \& Q \vdash P$ and $P \& Q \vdash Q$. Ascending to a linguistic plane of reference, we conclude that the truth of 'P' follows from the truth of 'P&Q', as does the truth of 'Q', and this gives us the three F entries in the truth-table for '&'. Similarly, putting 'P' for X and 'Q' for Y in &-I, we can conclude that the truth of 'P&Q' follows from the two-premiss set Z = ('P' is true, 'Q' is true), which gives us the T entry in the table.

On this view, the semantics of a connective is answerable to its rules of proof. So what function does the semantics play, if it is the inference rules which are fundamental? We can say that the semantics provides a tool for establishing that an argument schema is *incorrect* where, in this context, 'incorrect' means 'not establishable by the rules of proof'. That a schema is incorrect, in this sense, if there is a counterexample in the semantic sense, is established by a soundness proof for the semantics, and that every incorrect schema has a semantic counterexample is established by the completeness proof. But according to Michael Dummett, the position on which the semantics is answerable to the rules of proof rather than conversely

> . . . obliterates the distinction between a semantic notion of logical consequence . . . and a merely algebraic one . . . Semantic notions are framed in terms of concepts which are taken to have a direct relation to the use which is made of the sentences of the language . . . algebraic notions define a valuation as a purely mathematical object . . . which has no intrinsic connection with the use of sentences . . . It is quite

[15] Quine [1970, pp. 11–12].

impossible that it should be an utter illusion that semantic accounts of the logical constants supply an explanation of their meanings, and that such accounts have no more significance than a purely algebraic characterization of a logical system.[16]

However, we have the materials at hand to rebut this objection, at least in the case of our example, the connective '&': a distinction between semantic and algebraic clauses for '&' may be said to be manifested in the fact that the semantic account of '&' follows from the rules of proof together with the Quinean manipulations of the truth-predicate and the classical assumptions that each sentence is either true or false and not both. So the semantic account *is* intimately related to the use which is made of sentences in the language, even if we think of the fundamental facts about the use of connectives as being recorded in their rules of proof.

How does all this apply to the modal operators? To simplify matters, let us suppose '◇' to be introduced by definition, and concentrate on '□'. The elimination rule for '□' is straightforward:

□—E: if X ⊢ □A, then X ⊢ A.

However, it is more difficult to say what the introduction rule should be. Intuitively, the idea is that if A follows from a set of sentences all of which are necessary, then A is itself necessary. But what does 'necessary' mean here? An obvious suggestion is that it means 'has "□" as its main connective', which gives the rule:

□—I: if X ⊢ A and every sentence in X has '□' as its main connective, then X ⊢ □A.

One application of □—I establishes that '□□P' follows from '□P', and in fact the system defined by the two rules above is S4, so we cannot use the rules to establish the S5 thesis that '□—□—P' ('□◇P') follows from '—□—P'. However, if we decide to count '—□—P' as a necessary sentence, on the grounds that every occurrence of a sentential letter in it is within the scope of a '□', then the inference goes through and we obtain S5; that is, S5 is the system defined by the rules □—E, as above, and

□—I*: if X ⊢ A and for every sentence S in X, each sentence letter occurrence in S is within the scope of some occurrence of □, then X ⊢ □A.

And we can also obtain quantified S5 by replacing 'sentence' with

[16] Dummett [1978, p. 295].

'atomic predicate or relation symbol' in □–I*. Of course, it is not an *objection* to the proof-theoretic approach that it delivers different systems on different construals of 'necessary sentence', since possible worlds semantics also delivers different systems. Perhaps, indeed, the proof-theoretic approach has the advantage, delivering a narrower range of systems, which might on that account be regarded as the natural ones.

For modal operators, however, it is harder to meet Dummett's point that the proof-theoretic approach renders the possible worlds semantics indistinguishable from algebraic semantics, and thus fails to explain our intuition of naturalness in the former. For we cannot relate the possible worlds semantics to the use which is made of modal sentences by an argument using no more than the resources needed to derive the truth-table for '&' from its deduction rules: such manipulations with the truth predicate will not take us from expressions containing modal operators to expressions containing quantifiers over possible worlds. However, we can appeal here to the anti-realist thesis about possible worlds sentences, that the meaning of these sentences is the meaning which belongs to the modal sentences of which they are renderings; then we can employ Quine's principles to move from:

(7) '□P' is true

to

(8) □('P' is true)

and then apply the anti-realist thesis to (8) to obtain a sentence which has the meaning which (8) has,

(9) In every possible world, 'P' is true.[17]

There is no plausible claim which could substitute for the anti-realist thesis here which would permit the production of a clause from an algebraic definition of validity as an equally natural competitor to (9), so again, the natural semantics can be related to our use of modal sentences in a way that others cannot be.

[17] It may seem that (8) does not follow from (7) because a sentence which is actually used to express a necessary truth could have been used to express a contingent truth. However, we can regard the predicate 'true' as short for 'true-in-L' for a fixed language L and then say that the semantic properties of a language are the same in every world (so the Falsehood Principle is not applied to semantic vocabulary if languages need not be necessary existents). Thus if '□P' is true in L, there is no world where 'P' expresses a contingent truth *and* is a sentence of L. I take this point from Peacocke [1978, pp. 477–8].

A technical problem for the proof-theoretic approach is to show that possible worlds semantics agrees with it about the validity and invalidity of modal arguments. To do this, we should prove that any modal argument is correct according to the inference rules (say, for S5) iff its translation into possible worlds language is valid according to ordinary first-order semantics.[18] This result establishes that even though world quantifiers in L_w do not mean what quantifiers usually mean in first-order language, the difference is not sufficient to prevent us from employing our familiarity with first-order logic in assessing modal arguments. And, in fact, the argument which is needed here is fairly trivial.[19] However, there is a more substantial philosophical difficulty

[18] There is a minor technical complication here, since we have not been translating modal sentences into ordinary first-order sentences, but rather into sentences of a *two-sorted* first-order language. In a model for such a language, two domains are specified, one for the range of variables of the first sort and the other for that of the second sort. Each n-place predicate of the language also has its sort specified; that is, each place of the predicate is stipulated to be of the first or of the second sort, and an atomic sentence formed from such a predicate is true only if it has names of objects from the appropriate domain in the appropriate places. For further details, see Enderton [1972, pp. 277-86].

[19] More precisely, we have to show that if H is a set of L_m sentences and C is an L_m sentence, then there is a proof of C from H in a natural deduction system with ($\Box$E) and ($\Box$I*) iff there is a first-order proof of Trans(C) from Trans(H) in L_w. However, there is a complication with natural deduction, since a natural deduction proof in L_w might take 'detours' through L_w sentences which do not translate any L_m sentences, and it is therefore conceivable that some L_w sentence A follows from a set X of L_w sentences, where A and the members of X all translate L_m sentences, but that every proof of A from X in L_w proceeds through some L_w sentence which does not translate any L_m sentence, so that there is no corresponding L_m proof. In fact, this does not happen, but it is easier to prove the result we want by using the syntactic method of trees. See Hodges [1977, p. 321] for the tree rules for connectives in first-order logic for the empty domain. To adapt this system to first-order S5, S. G. Williams suggested to me the strategy of numbering all the nodes of a tree, the first node being numbered '0'. When a new node is generated by applying a rule to an earlier node, the new node has the same number as the earlier node, except if the rule applied is either of the following rules for the modal operators:

($\Box$): If a formula of the form $\Box$A occurs on a branch B at a node numbered m, then any branch B′ extending B may itself be extended by the addition of a node containing just the formula A, and this node may have any number.

($\Diamond$): If a formula of the form $\Diamond$A occurs on a branch at a node numbered m, then any branch B′ extending B may itself be extended by the addition of a node containing just the formula A, the new node to have a number which does not number a node already on B′.

In addition, there are the obvious shift rules (—$\Box$) and (—$\Diamond$), the quantifier rules from footnote 7 to Chapter 2, and one extra rule, which allows us at any stage in

in the way of the present approach. The thought that the rules of inference embody the meaning of a connective is supported by the idea that the rules explain what operation on meanings the new connective performs, so that if it is introduced into an already understood language, then one can straight away understand sentences with one occurrence of the new connective, since one grasps the operation (knowing the rules) and also grasps the meanings operated upon (since these are from the already-understood language in this case); and one's understanding of sentences with more than one occurrence of the new connective is built up from there. But inspection of the introduction rules for '□' given above reveals that they do not help in the step from a '□'-free language to sentences with one '□', since they only specify how to reason with '□' in a language to whose lexicon it already belongs. This suggests that the rules do not embody an operation on meanings of the required kind.

Perhaps this difficulty can be overcome recursively. But, at this point, it is simpler to introduce another approach to which an anti-realist might turn for his primary account of what it is for a modal argument to be valid. This second approach, which involves alternative semantics rather than proof theory, can be motivated by considering how we went about engaging intuitions about correctness and incorrectness of formal inference schemata in §2 of Chapter 1. Our procedure there was to choose particular English substitution-instances of the given schema such that, for an invalid schema, the possibility of the premisses of the instance being true while its conclusion is false, is completely evident. This method gives rise to a conception of validity for formal languages which is the main semantic rival to the usual

the construction of a tree to append to any branch a node with any number which contains any number of instances of the formula '$\Diamond E(t)$', where t is any name; this last rule reflects the fact that in the model theory each x in D is a member of some d(w). In this system, we say that a branch closes iff it contains two nodes with the same number such that A occurs at one and $-A$ at the other, or a single node at which A and $-A$ both occur, or a node at which some formula of the form $-(t=t)$ occurs. If we now use our translation procedure to go from L_m to L_w, and use the standard quantifier rules for '$\forall$' and '$\exists$' regardless of the sort of variable, there is an exact one-one correspondence between trees for modal arguments and trees for their possible worlds translations; in particular, if an L_w tree closes while there is no corresponding closed L_m tree, this is because some formula at the top of the tree, a premiss or the negation of the conclusion of the L_w argument, is not the translation of any L_m sentence. The reader may prove this by an inductive argument (take 'w*' to be the 0th world constant, and match applications of the quantifier rules in L_w to quantifiers binding world variables with node numbers in L_m trees).

model-theoretic approaches, the so-called *substitutional* conception, and it is applicable to the modal case.[20] The basic idea would be that a schema of quantified S5 is valid iff uniform substitution of expressions with n free variables for the n-place atomic relation symbols of the schema, and names for the individual constants, always yields arguments whose premisses cannot all be true if their conclusion is false. As it stands, this notion of validity is implicitly relativized to the language from which the *substituends* are chosen; at present, we have in mind a regimented fragment of a natural language. However, if the fragment is sufficiently weak, the wrong results will be obtained; for instance, if we restrict ourselves to arithmetical expressions and names of numbers, then we will validate the inference of '□P' from 'P', since all truths of arithmetic are necessary truths. A better definition is therefore that a schema is valid for L iff it is not possible that there is some extension of L from which substituends can be chosen in such a way that it is possible for the premisses of the resulting instance to be true while its conclusion is false.[21]

Several nice questions arise about how the modal substitutional account is to be understood. It is no objection to it that it uses modal operators to define validity for modal languages, unless it is also objectionable that quantifiers are used to define validity for quantificational languages. But since the definition speaks of what is possible and not possible for extensions of the language in question, a complete account would have to address itself to such topics as the existence conditions of languages: does a language exist at a world only if it is the actual language of some population at that world?[22] And, as with the previous case, we would like to show that a modal argument is valid by the substitutional account iff its translation into possible worlds discourse is valid by ordinary first-order semantics. However, if T(R) is the translation into L_w of some modal argument R, then by ordinary first-order semantics T(R) is valid iff T(R) has a proof. But from our discussion of the proof-theoretic approach we know how to transcribe a proof of T(R) back into modal language to obtain a proof of R (see footnote 19), which shows that the validity of T(R) in first-order terms implies the validity of R in substitutional terms provided our rules of proof for modal arguments are sound, where soundness is also conceived of

[20] See Quine [1970, pp.. 49–56]. [21] See Peacocke [1981, pp. 137–8].

[22] There is a detailed treatment of this and other questions about the substitutional account in the doctoral dissertation of S. G. Williams (Oxford University 1984).

substitutionally. For the converse, we would need to know that these rules are complete as well, but it is beyond the scope of this book to investigate soundness and completeness conceived substitutionally; in fact, soundness, the more important property, is easy to establish, but completeness is highly problematic.[23]

In sum, then, the anti-realist is by no means at a loss to explain what it is for modal arguments to be valid or invalid, even though he rejects possible worlds semantics as giving the fundamental account of these concepts for modal systems.[24]

5. *The meanings of possible worlds sentences*

The anti-realist thesis that the meanings of L_w sentences are imputed to them one by one by the inverse of the translation scheme which carries L_m sentences into L_w sentences figured in the previous discussion, as providing the anti-realist with a reply to Dummett's objection. But the thesis itself requires some defence. The problem is that there are L_w sentences which are apparently meaningful but which are not reverse-translated by any L_m sentences; nor can these sentences be eliminated merely by deleting some vocabulary from L_w, since some of the problematic L_w sentences employ only vocabulary which appears in L_w sentences which do translate L_m sentences, so that deletion would result in these L_m sentences having no possible worlds rendering. As an example of such a problematic L_w sentence, Hazen has given:

(10) $(\forall w)(\exists x)(E(x, w) \,\&\, E(x, w^*))$.[25]

[23] Here I am indebted to Williams (op. cit.).

[24] A certain kind of 'conservation' result is implied by the position just reached: if there is an inference in first-order possible worlds language which translates back into a modal inference, then it is valid only if the modal inference is valid. So deducibility in the extensional language is conservative with respect to deducibility in the modal language. Furthermore, in both languages, deducibility is equivalent to semantic consequence, so conservativeness holds for this relation too. But in a more general setting we lose the conservation property, in particular, if we allow second-order quantifiers into the extensional metalanguage, for then we can find examples of incomplete logics (though S5 is not one) where a certain inference in the metalanguage is second-order valid, and *also* has a second-order proof, but its reverse translation into modal language lacks a modal-logical proof. However, despite the formal interest of this topic, it is not relevant to our present concerns, since the modal consequence relation which causes the problem is not one at which one would arrive from the kind of starting point at which the question of the justification of possible worlds semantics arises. See van Benthem [1979] for further details.

[25] Hazen [1976, p. 38].

(10) says that in every possible world there is some object which also exists in the actual world, and cannot be reverse-translated into L_m because no expression of that language as it presently stands can have the force of 'x actually exists' if that phrase is within the scope of a modal operator.

However, (10) has a perfectly natural English rendering, 'Necessarily, some actual object exists', which does not contain any of the vocabulary of possible worlds. So what the example shows is that L_m as it stands does not have all the resources required to formalize English modal discourse. Hazen suggests that L_m be supplemented with an 'actuality' operator, written 'A', to obtain an expanded language (which we shall still call L_m) in which we can express (10), i.e., in which we can formalize 'Necessarily, something actual exists':

(11) $\Box(\exists x)\mathsf{A}(E(x))$.

A similar example involves the L_w sentence:

(12) $(\exists w)(\exists x)(E(x, w) \,\&\, -E(x, w^*))$.

(12) says that in some world there is something which does not exist in the actual world, which in modal English is 'there could have been things other than there actually are', a truth which underlies some of our counterexamples to Barcan and converse Barcan formulae in §2 of Chapter 2. To express this with the new operator, we write:

(13) $\Diamond(\exists x) - \mathsf{A}(E(x))$.

So far, the introduction of the actuality operator has merely been a syntactic manoeuvre. To ensure that (11) and (13) do have the readings (10) and (12) respectively, we need to give an evaluation clause for the new operator in possible worlds semantics. But there is no mystery about what the clause should be:

(i) a formula of the form $\mathsf{A}(B)$ holds at a world w in a model M iff B holds at the actual world w* of M.

The reader may confirm, by applying the relevant evaluation clauses to (11) and (13), that they do have the possible worlds imports stated in (10) and (12).[26]

However, having added 'A' to L_m, Hazen is still able to find problematic L_w sentences. For instance, if instead of a constant for the actual world in (10) we have another quantified world variable,

[26] For more about 'actually' see Davies and Humberstone [1980].

(14) $(\exists u)(\forall w)(\exists x)(E(x, w) \& E(x, u))$

then our simple 'actually' operator is of no avail. It is tempting to render (14) in English as 'it could have been that necessarily, something is actual', but in S5 this piece of English on its most natural interpretation is simply equivalent to 'necessarily, something is actual'. To obtain the effect of (14), we have to allow 'actual' to refer back to the state of affairs introduced by an evaluation of the initial 'it could have been that': we wish to speak of what would have been actual if that state of affairs had obtained. However, it is not unreasonable to hold that it is also possible to hear the English as involving such a back reference, and in that case we would be justified in introducing further operators into L_m to express this reading. Such operators, 'indexed' "actually" operators, have been devised by Christopher Peacocke.[27] The idea is to attach a numerical index to the '$\Diamond$' or '$\Box$' which we wish the later occurrence of 'actual' to pick up, and then to index the corresponding 'actually' operator with the same number. We can now give (15) as the L_m sentence which (14) translates:

(15) $\Diamond_1\Box(\exists x)A_1(E(x))$.

In possible worlds terms, indexed 'actually' operators are required when one wishes to assert that things meeting a certain condition at one world also meet some condition at some perhaps distinct world; the problem which the indices solve is that of expressing the condition met at the first world in such a way that in evaluating the condition at another world, one is taken back to that first world. Obviously, then, one cannot give an evaluation clause for indexed operators in the straightforward manner of (i) above, since in evaluating a formula one needs some method of keeping track of which worlds are introduced for which indexed "$\Box$"'s and "$\Diamond$"'s as one proceeds. However, the details are not too complicated.[28] Later, we shall see that these indexed

[27] See Peacocke [1978, pp. 485–7].

[28] To keep track of worlds in the course of an evaluation, we conduct evaluations with respect to an assignment of numerical indexes to worlds. Suppose we write '$s[w/i]$' for the assignment of indexes to worlds just like s except that $s[w/i](i) = w$, and in all our evaluation clauses of § 2 of Chapter 2 we replace 'is true at world w' with 'is true at world w with respect to s'. We then have the following clauses for '$\Diamond_i$' and 'A_i':

($\Diamond_i$): $\Diamond_i A$ is true at a world w with respect to s iff for some u in W, A is true at u with respect to $s[u/i]$;

(A_i): $A_i B$ is true at a world w with respect to s iff B is true at $s(i)$ with respect to s.

The definition of model for quantified S5 is not changed, and we say that a

operators are indispensable in the formulation of some plausible modal theses about entities of various sorts; and in §3 of Chapter 3 we already have a thesis which needs them for its exact expression, the thesis that there are infinitely many non-existents at each world.[29]

Further generalization of the basic 'actually' operator 'A' can be motivated by consideration of such a sentence as

(16) My car (a) could have been the same colour as yours (b) actually is.[30]

In possible worlds terminology, (16) says that in some world my car has the same colour as the colour which yours has in the actual world, and thus involves a *cross*world comparison of the colour of two objects. One could give a formalization of (16) by quantifying over colours:

(17) $(\exists C)((\Diamond(a \text{ has } C)) \,\&\, b \text{ has } C)$.

But (17) quantifies over abstract objects, colours, and therefore does not sit well with the anti-realist approach we are at present expounding, since this approach is motivated in part by nominalism, which denies that there are abstract objects of any sort. However, we can give a nominalist formalization of (16) in two steps. First, even a non-modal sentence can involve apparent quantification over colours, for instance, 'my car is the same colour as yours'. To formalize this without quantifying over colours, we need a two-place predicate C so that we can write 'C(a, b)' for 'a is samecoloured with b'. In possible worlds language, we now wish to be able to compare a in a world u with b in a distinct world v, so that we can say that a in u is samecoloured with b in v.[31] That is, we require C to be a *four*-place predicate in L_w; then (16) has the L_w regimentation:

(18) $(\exists w)Cawbw^*$.

sentence is true in a model iff it is true at the actual world of the model with respect to some assignment s. It is then easy to see how the tree method of footnote 19 can be extended to give a sound and complete positive test for validity in L_m with indexed 'actually' operators and the new evaluation clauses.

[29] The formula we want is in a modal language with the appropriate operators which is based on the infinitary language $L_{\omega_1\omega_1}$. We write:

$$\Box_1 \exists\Diamond\{x_i : i \in N\}[\bigwedge(-A_1(E(x_i))) \,\&\, \bigwedge\{x_i \neq x_j : i < j, i, j \in N\}].$$

Here '$\exists\Diamond X$' abbreviates an infinite sequence of alternated "$\Diamond$" 's and existential quantifiers, one for each variable in X, the sequence beginning with '$\Diamond$'.

[30] The example is due to Christopher Peacocke.

[31] For further discussion of the crossworld predications, see Salmon [1981, Chapter 13].

To achieve the effect of (18) in L_m, one needs an 'actually' operator which can associate particular objects with worlds introduced by evaluations of modal operators, independently of the order in which the objects are referred to in the sentence. So we need numerical indexes on modal operators and 'actually' operators as before, but an 'actually' operator which is to associate a particular object with a given indexed '$\Box$' or '$\Diamond$' should be indexed with the name of that object as well as the index on the modal operator.[32] By allowing such double indexing in L_m, we can express (18) by

(19) $\Diamond_1 A_1^a A^b Cab$;

that is, on the appropriate method of translating L_m with doubly indexed 'actually' operators, (19) is translated as (18). Moreover, if we replace 'w*' in (18) with an existentially quantified variable, repeating the manoeuvre which took us from (10) to (14), we obtain:

(20) $(\exists u)(\exists w)Cawbu$

which is the translation in L_w of an L_m sentence with two doubly indexed 'actually' operators,

(21) $\Diamond_1 \Diamond_2 A_1^a A_2^b Cab$.

Evidently, it will be even less straightforward to add evaluation clauses for doubly indexed operators, and the translation scheme for L_m into L_w will also need modification, since we are now translating n-place L_m predicates by $n+2$-place L_w predicates, which is a new departure. But, once more, the technical details are not too difficult to work out.[33]

Should the anti-realist object to these new operators, on the grounds that they are really nothing but devices for disguised quantification

[32] This is a simplification of a notation proposed by Peacocke.

[33] Not much illumination is gained from spelling out the details, but for maximum generality we might want to consider relations $D^* \times W \times \ldots \times D^* \times W$, where '$D^*$' stands for a product of D by itself n times, for variable n in its different occurrences in the above expression. The object-language effect is obtained by indexed "$\Box$"'s and "$\Diamond$"'s and co-indexed 'actually' operators with n superscripts, the value of n being that of the corresponding $\star$. Extensions of appropriate types would be assigned to L_m predicate symbols, the intraworld and crossworld use of the predicate being unified by treating the intraworld use as a special case of the crossworld use when the worlds across which the predication is made are the same. For instance, the extension of 'is the same colour as' at w is given by the sequences of 4-tuples of the form $\langle x, w, y, w \rangle$ which belong to the general extension, which itself is a set of 4-tuples of the form $\langle x, u, y, v \rangle$. Strictly, this strategy requires us to find 'suppressed' indexed 'actually' operators when intraworld predication is made, the indexes being redundant in this case.

over worlds? The objection appears doubtful, for each successive step in introducing the operators was motivated by the production of an *English* sentence which required, or had a reading which required, the operator introduced at that step. And the English examples used no more give the appearance of quantification over worlds than do modal sentences of English formalizable in the original version of L_m, without any of the new operators, where the anti-realist *denies* that there is quantification over worlds. Thus it is unclear why he should think that the new operators import the unwanted ontology.

By introducing the various types of 'actually' operator, the anti-realist can assign interpretations to a much wider range of L_w sentences than is possible with L_m as it first stood. Nevertheless, there are still some L_w sentences which are uninterpreted, and which will remain so no matter how we might extend the techniques of the previous paragraphs. These are the sentences which are, intuitively, 'about' worlds, such sentences as '$(\forall w)(w = w)$', that is, 'every world is identical to itself'. Now it is possible for the anti-realist to rule such sentences out of L by restrictions on the methods by which well formed formulae can be built up,[34] but the fact remains that in formulating possible worlds model theory, especially if an accessibility relation is involved, one makes stipulations which, if formalized in first-order language, would be L_w sentences with no L_m interpretation.[35] So given that such sentences are used, what attitude should the anti-realist take to them? The most appealing suggestion is that he should regard them in the same way as certain mathematical sentences were regarded by Hilbert, that is, as instruments or uninterpreted stipulations which enable us to establish facts of interest about the interpreted sentences.[36] For the anti-realist, the justification for the use of such sentences, that is, for the making of such stipulations, is that a semantical theory conforming to them is in agreement, over questions of validity, with the fundamental account of validity in L_m, whichever account he prefers; the

[34] For example, rather than regard our two-sorted L_m as a notational convenience for a single-sorted language with predicates for being a world and being an individual, as is usually done (so that '$(\forall w)(w = w)$' is regarded as an abbreviation of '$(\forall x)(Wx \rightarrow (x = x))$'), we can think of the sort-distinctions as essential to the definition of wff in L_w; then if we stipulate that '=' is a predicate of sort $\langle D, D \rangle$ only, '$(\forall w)(w = w)$' will not even be well-formed.

[35] Mention of the accessibility relation here raises the broader issue of what first-order properties of this relation can be expressed by modal formulae. See Goldblatt [1975] and van Benthem [1979] for more on this question.

[36] See Dummett [1978, p. 219], Smorynski [1977, pp. 822–5], and Field [1980] for more details about Hilbert.

function of the uninterpreted sentences is just to ensure that this agreement obtains.

Finally, we should recall that the anti-realist treatment of L_w sentences was originally premised upon the thought that 'Possibly, P' and 'There are some possible circumstances in which P' are synonymous, the meaning of the former fixing that of the latter. But in our discussion of anti-realism, we have replaced possible circumstances by possible worlds, and it is a fair complaint that this detracts from the plausibility of the synonymy claim. For possible worlds are complete ways things might have been, and there is apparently nothing in the meaning of 'Possibly, P' which corresponds to this element of completeness. However, in Chapters 1 and 2 we developed the materials for an anti-realist response to this objection, for we showed there how to give a semantics for L_m which preserves the quantifier treatment of '$\Box$' and '$\Diamond$' but does without the completeness assumption. The premise upon which the anti-realist treatment really relies might be better expressed as postulating a synonymy between 'Possibly, P' and 'There is some possibility that P', where, again, the latter is a mere paraphrase of the former. We already know that possibility semantics is equivalent to possible worlds semantics, and so all the apparatus of this and the previous section can be transferred to that system.[37] Possible worlds are complete possibilities, that is, possibilities which have no proper refinements, and we can express this condition in the language of possibility semantics. Again, the resulting sentence will have no L_m interpretation, since it is 'about' possibilities, but the stipulation that every possibility conform to the condition is an example of an uninterpreted sentence which is of considerable instrumental value, since it allows us to replace the clause for negation which is the cause of the complexities in possibility semantics and obtain the familiar possible worlds semantics as a result.

So we may conclude that anti-realism about possible worlds does not demand the wholesale rejection of the semantical methods of the earlier chapters, and thus we can continue to use these methods so long as we find them useful, without having to believe that there literally are such things as worlds.

[37] For reverse-translating L_w into L_m, see §3 of the Appendix. In reverse-translating a sentence about possibilities, when a possibility quantifier '$(\exists X)$' is eliminated, so are all occurrences of '$-(\exists Y \geqslant X)$', '$-$' replaces this expression in front of the complete subformula within the latter's scope, and 'Y' is eliminated from this subformula.

5

A Modal Theory: The Essences of Sets

1. *Essential properties and essences*

THE study of non-modal predicate calculus is often labelled 'mathematical logic', a title which adverts to one motivation for the development of modern logic, which was to codify the canons of inference employed by mathematicians in constructing and developing mathematical theories. By analogy, one might expect to find in modal logic the canons of inference employed in the construction and development of modal theories. However, modal theories themselves, at least for the entities which will be of interest to us, are in a somewhat underdeveloped state in comparison with mathematical theories, with but one exception;[1] we will therefore be studying individual modal theses rather than full theories. Moreover, the subject-matter of these theses, typically philosophical, is not very amenable to illuminating formal articulation (again with that exception), so we will not be treating the theses as axioms and trying to prove theorems from them. The role of modal logic is more to make the theses absolutely precise than to facilitate the deduction of substantial consequences from them; it also helps provide a framework for clear discussion of the grounds for accepting or rejecting these theses. It is this activity which we will pursue in this and the next three chapters, with a view to meeting Quine's challenge to explain transworld identity. For this challenge does not evaporate upon adoption of anti-realism about worlds; as we shall see, an elucidation of transworld identity can be regarded as an elucidation of the

[1] The exception is modal set theory, for which see Fine [1981b]. There is another sense in which the modal *logics* we surveyed in Chapter 1 might be regarded as theories, in that the logic determined by the class of general models, usually known as K, could be taken to be the basic logic, and other (adequately axiomatizable) logics thought of as theories expressing in modal terms the fact that accessibility has such-and-such a structure. But this viewpoint does not sit well with the anti-realist attitude to possible worlds semantics.

boundaries of possibility for ordinary things, an enterprise which an anti-realist about worlds must take seriously if he thinks there are objective modal facts, even if they are not about worlds.

The modal theses with which we will be concerned are theses attributing essential properties and individual essences to things of given categories, so we start with an explanation of this terminology. First, an *essential property* of an object x is a property without possessing which x could not exist. In other words, if P is an essential property of x, then for all possible worlds w, if x exists in w then x has P in w. Essential properties may be as complex as you please, but there are also simple and highly trivial examples of them. For instance, the property of *existing* is obviously essential to every object x, regardless of what kind of a thing x is. The same is true of the property of being *self-identical.* However, as we have set up our system of quantified S5, there is a difference between these two properties, for the latter is also a *necessary* property of x. We may say that a necessary property is a property possessed by x in every world (not just in worlds where x exists). Then for material objects, the paradigm of contingent existents, existence will not be a necessary property, but self-identity will be, since even if an object a does not exist at a world w, our semantics makes 'a = a' true at w. However, having drawn this contrast between necessary and essential properties, we will not refer to it again, since from now on we will be concerned only with essential properties.

The essential properties of a thing will typically depend upon what category of thing it is, and perhaps also on some more particular facts about the thing itself, so a thesis attributing an essential property will say, roughly, that for things of such-and-such a category, if certain particular facts about them are thus-and-so, this remains the case in every world where the thing exists. In possible worlds terminology, what we want to say is that for any possible object which is of category C in some possible world, if that thing has a certain (perhaps complex) property P at that world, then at every world where that thing exists, it still has P. As a modal schema, this possible worlds formulation translates into:

$$\text{(S)}\quad \Box(\forall v)\Box[(C(v) \,\&\, A(v)) \rightarrow \Box(E(v) \rightarrow A(v))].$$

Note that (S) begins with a universal quantifier governed by a '□', so instances of (S) will speak of every object in every possible world, that is, all possible objects. The point is that one does not wish to limit the attribution of an essential property to just the things of a given kind which happen to exist. The schematic letter 'C' stands for a predicate which identifies the category of thing to whose members we wish to attribute the essential property; sample instances of C are 'number', 'set', 'organism', and 'event'.[2] What follows in A(v) specifies a particular property which the consequent of the main conditional asserts to be essential. If we write out the trivial instance of (S) which says that existence is an essential property of every possible object, we obtain:

$$(1)\quad \Box(\forall x)\Box[(T(x) \,\&\, E(x)) \rightarrow \Box(E(x) \rightarrow (E(x)))].$$

Here 'T(x)' stands for 'x is a thing'; so in effect there is no restriction on the category of thing to which we wish to attribute existence as an essential property, and we could just as well have suppressed the symbols '(T(x)&' in (1). In possible worlds terms, (1) says that for all possible x, if x is a thing and exists in some world, then in any world in which x exists, it exists. This way of reading (1) is justified by the equivalence of (1) with

$$(2)\quad \Box(\forall x)[\Diamond(T(x) \,\&\, E(x)) \rightarrow \Box(E(x) \rightarrow E(x))],$$

which brings out the function of the second '□' in (S). In future, when the context makes it clear what category of object it is which we are talking about, we suppress the category predicate. Of course, in the examples to come, the formula A(v) will be more complicated and more interesting than 'E(x)', and is therefore likely to contain variables other than v bound by quantifiers following the initial '∀v'. So (S) is really a simplified version of a general schema for an attribution of an essential property to things of a category C. The general schema is:

$$(S')\quad \Box(\forall v)\Box(\forall u_1)\ldots\Box(\forall u_n)\Box[(C(v) \,\&\, A(v, u_1 \ldots u_n)) \rightarrow \Box(E(v) \rightarrow A(v, u_1 \ldots u_n))].$$

The reader who finds the point of this schema difficult to grasp in the abstract should return to it once we have presented its first substantial instance, the thesis of Membership Rigidity for sets, later in this chapter.

So much for the idea of an essential property. The other notion we mentioned as an object of investigation was that of an *individual essence*, and this concept can be defined in terms of essential properties.

[2] My use of 'category' is not governed by any underlying philosophical theory of categories. Roughly, I distinguish categories when there are interestingly different things to be said about individual essence.

An individual essence of an object x is a set of properties I which satisfies the following two conditions:

I(i) every property P in I is an essential property of x;
I(ii) it is not possible that some object y distinct from x has every member of I.

Note that in possible worlds terminology, I(ii) says that there is no world w such that some y other than x has every P in I at w. As with essential properties, there are trivial examples of individual essences, of which the most obvious involves self-identity: the property of being identical to a is an essential property of a and furthermore, it is not possible that any object other than a has that property. This fact suggests that we might define a special sub-class of individual essences, which we can call *non-trivial* individual essences. We say that a non-trivial essential property of x is any property essential to x other than:

(a) a property x has as a consequence of some *de dicto* truth (this excludes, e.g., the property of being unmarried if a bachelor);
(b) the properties of existence, self-identity, or their cognates (by the 'cognate' of a property P, let us mean any Q such that if x has P, then it follows by logic alone that x has Q; thus, for the number three, being identical to three and being identical either to three or to four, are both trivial essential properties, while being the cube root of twenty-seven is a non-trivial essential property, since the fact that three is the cube root of twenty-seven follows only in theories containing some portion of elementary arithmetic);
(c) a property x has in virtue of a necessary truth concerning items of another category (so it is only trivially essential to material things that they are such that three is the cube root of twenty-seven; the criterion also makes it trivially essential to an object x which stands in some necessary relation R to a thing which is necessarily F that it is necessarily R to some F, so it does not quite capture the idea of irrelevance or independence which underlies the example given; it would take us too far afield to improve this here).

Finally, we define a non-trivial individual essence of x to be an individual essence of x none of whose members are trivial essential peoperties.

From now on, our only interest will be in the non-trivial so, unless explicitly stated otherwise, this qualification will be understood whenever the phrases 'essential property' or 'individual essence' are used. The motivation for investigating individual essences should be obvious, since if every object has such an essence, the problem of elucidating transworld identity can be solved. It will be recalled from §2 of Chapter 3 that, according to Quine, there are relatively unproblematic criteria of cross-moment identification of bodies, but no such criteria of crossworld identification. However, an individual essence of an object x, in virtue of its non-triviality, would give necessary and sufficient conditions for crossworld identification of x without employing the property of being identical to x or any of its cognates. Each property in x's essence would correspond to a necessary condition, since each of them is essential to x, and the whole set of properties would give a sufficient condition, because of clause I(ii) (of course, we shall hope to find rather simpler conditions than those resulting from taking the *whole* set). The resulting transworld identity condition for x, as already remarked, would apply only to pairs of worlds u and v such that x exists in both, but this is as complete a solution to the problem as Quine's criteria of continuity of displacement, distortion and chemical change are to the corresponding problem of transtemporal identity. So our plan is to choose particular categories of object and develop a theory of the individual essences of members of the given category; we can also describe our theory as a theory of transworld identity conditions, but only as a *façon de parler*, since we are anti-realists about worlds. We will begin with that category of object, sets, for which the correct account of individual essence is most obvious.

2. *The essences of sets*

In this section, we are going to develop some modal intuitions about sets, with a view to formulating, and then defending, certain claims about the essences of sets; we shall argue that the members of a set yield an essence for the set, in the sense that if certain objects are the members of a set in some world, then the property of having exactly those objects as members is an essence of that set. Evidently, this claim can be broken apart into a necessary and a sufficient condition for transworld identity of sets, the condition of having the same members; we shall see below exactly how to say in modal language that this condition is both necessary and sufficient.

There are some preliminary matters to which we should attend. First, exactly what is a set? In what follows, we shall presume a certain conception of set known as the *iterative* conception, which is best explained by contrast with another conception, according to which a set is any collection of entities which fall under some particular concept. As is well known, the unqualified version of the latter view was shown by Russell to be inconsistent. What Russell did was to find a concept c such that the 'set' which comprises those things which fall under c can be shown to be a logically impossible object; hence no set is specified by this concept, and thus the view of sets in question is refuted. The concept Russell began with is the concept of being self-membered, a concept which specifies the set of all objects which are members of themselves. If we use '$\in$' for the relation of membership, this set may be designated:

$$\{x: x \in x\}$$

(read: the set of all sets x such that x is a member of x). Now this appears to be a quite reasonable specification of a set; for example, the set of all abstract objects is itself an abstract object, and is therefore a member of itself. But if we next consider the concept of not being self-membered, which specifies the Russell set R designated by

$$\{x: -(x \in x)\}$$

we can derive a contradiction by asking whether or not R is a member of itself: if R is a member of R, then by the definition of R it is not self-membered, i.e. it is not a member of R, while if R is not a member of R, then by the definition of R it is a member of itself, i.e. a member of R. That is, R is a member of R iff it is not a member of R; but in classical logic, '$P \leftrightarrow -P$' is a contradiction.

This argument shows that the conception of set as the range of items falling under a concept must be revised, and a very natural revision is suggested by the argument itself. For, on reflection, there is something strange about the idea that a set could be a member of itself. To see this, take for an analogy the idea that God is self-created. The problem with this doctrine is that for 'x creates y' to be true, x must bring about the existence of y, and bringing about is an activity the indulgence in which by x requires the existence of x. Thus God's existence is a precondition of His creating anything, including Himself: He cannot create Himself without satisfying a precondition which renders His creating Himself unnecessary, indeed, impossible. In a parallel fashion, it seems that a set cannot be a member of itself if we think of a set as a collection of entities which are somehow 'brought together', the set existing only when these entities *are* brought together. Then if a set was one of its own members, it would have to pre-exist itself, in order to be amongst the things whose being brought together constitutes the formation of the set.

These arguments against self-creation and self-membership evidently play upon some 'before/after' dichotomy. In the case of self-creation, the dichotomy arises from the *a priori* truth that the existence of the creator must temporally precede that of the creation. In the case of sets, the use of temporal language is unavoidable, but metaphorical: there is of course no *physical* process of bringing together certain antecedently existing objects, a process whose completion at a certain point in time marks the start of the existence of the set. But however hard it may be to explain the appropriateness of the metaphor, its naturalness cannot be gainsaid, and it motivates a picture of the universe of sets something like the following. At the bottom level, we have all the non-sets, for which the term 'individual' is reserved. At the next level, all sets whose members are entities at the bottom level, i.e. all sets of individuals, are formed. At the level after that, all sets whose members are objects on one or other of the previous two levels are formed; at the level after that, all sets of objects from the previous three levels are formed, and so on for every finite level. The first infinite level is reached by forming all sets whose members are objects to be found somewhere on the finite levels, and then the process is iterated until the level of the next infinite limit ordinal, and then iterated again, and so on, without end. This sequence of levels is known as the *cumulative hierarchy* of sets. It is cumulative because at any given stage one may at that stage form any set so long as each member of the set appears at *some* previous stage—one is not limited to objects from the immediately preceding stage. And because a set is formed only at a level after all its members have been formed, the analogy between set-formation and object-creation is preserved: no set in this hierarchy is a member of itself.[3]

The conception of sets which the cumulative hierarchy embodies is the iterative conception, which has been extensively studied in the form of axiomatic theories: axioms are laid down intended to express the fundamental principles of the iterative conception, and their

[3] There is a careful investigation of the relationship between the theory of levels, or stages, and the axioms of ZF, in Boolos [1971], where the author gives a metaphorical account of forming a set, which he attributes to Kripke, as putting a 'lassoo' around its members (op. cit., p. 220, fn. 7).

consequences are investigated, along with metamathematical properties of the axiomatic theory itself. The best-known of all these theories is the Zermelo–Fraenkel theory of pure sets, usually called ZF; the sets are said to be pure because the ground level of the hierarchy, the level of individuals, is taken to be empty. Thus the hierarchy begins only with the second level, where we form all sets of individuals. Since there are no individuals, the set of all individuals is the empty set; indeed, *any* set of individuals is the empty set, so the empty set is the only set at the second level, but the hierarchy of pure sets can be built up from there.[4]

However, the assumption that there are no individuals is rather special, and there are versions of set theory in which it is not made. Such a theory is ZFI, Zermelo–Fraenkel set theory with 'individuals' (sometimes called 'ur-elements'), to which we shall advert further in §3 of this chapter. But one device of ZFI is worth noting at this point. ZFI is a theory of two kinds of entities, individuals and sets, and some of its assertions are not assertions about all *objects*, but only about all sets, or all individuals (nothing is both). Consider, for instance, the so-called Axiom of Extensionality, which in fact is a principle common to all reasonable conceptions of set. This axiom says that sets with the same members are the same set or, equivalently, that if x and y are distinct sets, then either x has a member which is not in y, or y has a member which is not in x; so the identity of a set is fixed by its membership. However, if we write the Axiom of Extensionality as follows:

(E*) $(\forall x)(\forall y)[(\forall z)(z \in x \leftrightarrow z \in y) \rightarrow x = y]$

we assert that for any *objects* x and y, if the members of x are the same as the members of y, then x and y are the same. In ZFI, the objects comprise both individuals and sets, and individuals have no members. Thus if x and y are distinct individuals, (E*) in fact identifies them, since, as individuals, they have the same members, none at all. What we wanted to say was only that if x and y are *sets*, then they are the same if they have the same members, so we need a predicate 'S' for being a set (and, for other assertions, a predicate 'I' for being an individual) to formulate the axiom as we should: (E*) should have begun '$(\forall x)(\forall y)(Sx \,\&\, Sy \rightarrow$'. However, a notational convention enables us to avoid expansion into longer formulae with extra implication signs; instead of using the predicates 'S' and 'I', we shall reserve

[4] See Devlin [1979] for a very accessible introductory account.

distinctive variables for sets and distinctive variables for individuals, using standard variables 'x', 'y', etc., as general variables, to range over both individuals and sets. For set variables, we shall use 'X', 'Y', etc., and for individual variables, 'i', 'j', etc. Formulae with the special variables may be regarded as abbreviations of formulae with the special predicates 'S' and 'I'. So we can rewrite the Axiom of Extensionality as

(E) $(\forall X)(\forall Y)[(\forall z)(z \in X \leftrightarrow z \in Y) \rightarrow X = Y]$.

(E) says that for any set X and any set Y, if they have the same members, they are the same set. We use the general variable 'z' for the members, since the members of a set may be either sets or individuals. Apart from this use of notation, however, no familarity with any set theory is needed to follow the philosophical discussion below and in §4, only a grasp of the conception of sets just outlined; therefore we will not further elaborate the details of ZF or ZFI here.[5]

The question to which we now wish to develop an answer is: what are the transworld identity conditions of sets? The answer we are going to give is the following:

> sets x and y in different worlds u and v respectively are the same set if and only if the members of x at u are the same as the members of y at v;

but we would like to approach this answer by a route which highlights its plausibility, and we would also like to see how it might be formulated in modal language as a thesis about the essences of sets. So let us work with a concrete case. Consider the set whose members are exactly the passengers on a certain transatlantic flight, say British Airways flight BA 167 from London to New Orleans on Sunday 19th September 1982. This set, which we will call X, is a set of people which we have picked out by using a characteristic possessed by all and only those people, the characteristic of being a passenger on that flight. The set itself, however, is merely the collection of those people. This feature of sets, that they are 'nothing but' collections of objects, is used by set-theorists to motivate the Axiom of Extensionality, (E), just given above. One might say that (E) expresses at least part of the idea that the identity of a set is completely given by the identities of its members, for if there were more than this to the identity of a set, there would be no reason to expect sameness of membership to yield a

[5] For ZFI, I have followed Suppes [1972].

sufficient condition for sameness of set. By contrast, if one thinks of the *characteristics* we can use to pick out sets, it would be a mistake to hold that it is sufficient for these characteristics to be the same that they apply to the same objects; for instance, even if the passengers on the flight mentioned are all and only the members of a certain religious sect, the characteristics, or properties, of being on that flight and of being a member of that sect, are still distinct. There is more than one way of bringing out the difference, but the simplest is to refer to the possibility that the sect members travel on a different flight, or non-sect-members travel on that flight with them; that is, there are worlds where the extensions of the properties are different, so they cannot be the same property.

If these comments about (E) are accurate, then (E) is no mere contingent generalization about sets; rather, it is a necessary truth about them, in the 'broadly logical' sense of 'necessary', in which case we may strengthen (E) to the Axiom of Necessary Extensionality, which we abbreviate (□E):

$$(\Box E) \quad \Box(\forall X)(\forall Y)\,[((\forall z)(z \in X \leftrightarrow z \in Y)) \rightarrow X = Y].$$

Now it is tempting to think that with (□E), we are on the road to formulating in modal language our idea that for sets to be the same across worlds is for them to share, transworld, the same members. But, in fact, (□E) does not say anything of the kind. (□E) is a *de dicto* principle which says that (E) is true at every world w, which in turn means that for every w, sameness of membership at w is sufficient for identity of sets *at w*; this is an *intra*world identity condition which gives us no clue as to when a set x existing at a world u is identical to a set y existing at a distinct world v, and is quite consistent with x and y being the same set even if x's members at u are not the same as y's members at v. In other words, (□E) is consistent with a set being like a box, into which different things can be put at different times (Fine's simile); and this is exactly what we want to rule out.

The most obvious strategy for 'improving' (□E) in this respect is to strengthen it to a *de re* principle by inserting further operators. So let us consider the simplest *de re* extension of (□E),

$$(3) \quad \Box(\forall X)(\forall Y)\Box\,[((\forall z)(z \in X \leftrightarrow z \in Y)) \rightarrow X = Y].$$

(3) says that for any compossible sets X and Y, if there is some world where X and Y have the same members *from amongst the existents of that world* (remember that '∀' is given the actualist interpretation)

then X and Y are the same set. However, it is not too difficult to see that this is false; for if X and Y are sets existing at a world u where they have different members from amongst the existents of u, and hence are different sets, we can take a world v where no member of X exists and no member of Y exists, so that X and Y do have the same members from amongst the existents of v, and are therefore the same set, by (3). But, as a matter of modal logic, X and Y cannot be distinct at u and identical at v; so the sufficient condition for identity in (3) is mistaken.

For the sake of definiteness in the discussion to come, it is as well to decide now an issue which has been ignored to date, that of whether or not the Falsehood Principle should be applied to the set-membership predicate '$\in$': should we say that a sentence such as '$a \in b$' is false at a world w if either a or b does not exist at w? In fact, there is good reason to favour imposition of the Principle, a reason which flows from the iterative conception and the picture of the cumulative hierarchy of levels we sketched above. If we ask what it is to be a member of a set at a world, in terms of the iterative conception, then the following answer is the most plausible: to be a member of a set at a world is to be one of the objects on which the formation operation was brought to bear when the set was formed, on the relevant level, at that world. Now forming a set is just what it is to bring it into existence, and thus if b has members at a world, it has been formed at that world and so exists at that world ('forming' applies to all members at once, rather than one by one). This means that the truth of '$a \in b$' should imply the existence of b. And so far as the members are concerned it would be very strange to hold that an existent might be brought into existence by an operation on non-existents (if this were possible, perhaps a non-existent God could bring himself into existence); so the existence of a seems to be implied as well. The upshot of these remarks is that imposition of the Falsehood Principle on '$\in$' is well-motivated, so we take the following thesis (F) to be an axiom about sets:

(F) $\Box(\forall x)\Box(\forall X)\Box(x \in X \rightarrow E(x) \,\&\, E(X))$.[6]

[6] It is a consequence of (F) that the assertion

$$\text{Socrates} \in \{\text{Socrates}\}$$

is false at worlds where Socrates does not exist. This may look strange, but it is an appearance one easily learns to live with. A quite distinct question (discussed at greater length in the next section) is that of what set, if any, the set abstract '{Socrates}' should be said to denote at w if Socrates does not exist at w. Since names of non-existents are allowable, the abstract could be said to denote singleton-

The reader can easily determine that the imposition of (F) does not affect how things stand with (3), so we are still looking for the right way to formulate transworld identity conditions for sets. Let us approach the question by another route, *via* our definitions of essential properties and individual essences; for, as we saw in §1, a theory of individual essence provides transworld identity conditions. Our first step, therefore, is to ask what plausible candidates there are for non-trivial essential properties of sets. Returning to our original example of the set X of passengers on BA flight 167/9.19.82, let us compare X with the set Y of passengers on the same flight (or 'flight-type') for 9.21.82, the following Tuesday. In view of the airline's later decision to discontinue this route, we can imagine that X's members are just the travellers a, b, and c, while Y's members are just the travellers d, e, and f. By (E), X and Y are different sets. However, it is contingent that these six people travelled when they did, so we can imagine a world u where a, b, and c are the passengers on BA flight 167/9.21.82, while d, e, and f are the passengers on 167/9.19.82. Since it is the transworld identity of sets, and not flights, which is in question at the moment, let us assume that flights are the same in two worlds iff they are by the same airline, along the same route, and at the same times. Note that this is a genuinely non-trivial assumption about the transworld identity conditions of flights, even although it presumes the transworld identity conditions of other entities, airlines, routes and times. Let us also introduce the labels 'W' and 'Z' for the respective sets of travellers in u on BA 167/9.19.82 and BA 167/9.21.82; so W's members are d, e, and f, while Z's members are a, b, and c. We are now using 'X', 'Y', 'W', and 'Z' as labels for sets picked out at a particular world, X and Y at w* and W and Z at u. Setting this out in detail, we have:

(i) X = the set of travellers in w* on BA 167 for the 19th, = {a, b, c};

(ii) Y = the set of travellers in w* on BA 167 for the 21st, = {d, e, f};

(iii) W = the set of travellers in u on BA 167 for the 19th, = {d, e, f};

(iv) Z = the set of travellers in u on BA 167 for the 21st, = {a, b, c};

Socrates at all worlds, i.e. it could be a rigid designator. But if one reads the abstract as 'the set of all existent objects x such that x = Socrates', '{Socrates}' will be a non-rigid designator, since it will denote the empty set at worlds where Socrates does not exist. We will in fact adopt this latter, actualist, reading.

In the circumstances described, there are then essentially two different ways of making crossworld identifications between the sets of travellers X and Y in w* and the sets of travellers W and Z in u. We can say (A) X = W and Y = Z, or (B) X = Z and Y = W. Let us consider (A).

To hold that X = W and Y = Z, as we already remarked, is to hold that sets are like boxes; just as we can put different things into the same box at different times, so we can have different objects in the same set at different worlds. However, although this is compatible with (□E), there is a strong intuition that (A) is the wrong option and that sets are not like boxes; this would seem to be part of the intuition that there is nothing more to the identity of a set than the identities of its members, a part not captured by (□E). In support of (A), we can at best cite the fact that:

(4) The set of travellers on BA flight 167/9.19.82 could have been {d, e, f}

from which we may be tempted to infer that

(5) {a, b, c} could have had d, e, and f as its members instead of a, b, and c.[7]

Since it is in virtue of u and worlds like it that (4) is true, from (4) we can conclude that X = W. The objection to this argument, of course, is that (4) is ambiguous in the way that all subjunctive English sentences with definite descriptions are.[8] In this particular case, the ambiguity is between the readings (6) and (7):

(6) It is possible that the set of travellers on BA 167/9.19.82 = {d, e, f};

(7) Concerning the set of travellers on BA 167/9.19.82, possibly that set is identical to {d, e, f}.

[7] At this point, it does not matter whether we read set abstracts as rigid or non-rigid designators of sets (see previous note), since (5) is false even on the non-rigid reading, because each of a, b, and c is necessarily non-identical with each of d, e, and f. Note, however, that on the actualist reading of abstracts to be given in § 3 below, the sentence

$$\Diamond(\{a, b, c\} = \{d, e, f\})$$

is in fact *true* if there is a world at which none of a, b, c, d, e, and f exist. For then at that world, each abstract denotes the empty set of that world. But none of this is incompatible with the falsity of (7).

[8] The problem is one of the relative scopes of modal operators and definite descriptions. For a classic discussion of the problem against the background of Russell's treatment of descriptions, see Smullyan [1948].

In possible worlds terms, (6) says that in some world there is a set which satisfies the given description in that world and whose members are d, e, and f in that world. This is clearly true, in virtue of u, but is of no relevance to the question whether or not X = W, since in (6) the description 'the set of travellers on BA 167/9.19.82' does not pick out X, but rather, whatever set satisfies the description in a verifying world with respect to which the phrase 'it is possible that' is evaluated. Thus neither (7) nor (5) follows from (6). On the other hand, although (7) entails (5) and supports the claim that X = W (because the 'concerning the F, . . . that F . . .' construction allows us to pick out X with the definite description 'the set of travellers on BA 167/9.19.82' and say of X that its members could have been d, e, and f), the mere existence of such a world as u does not suffice for the truth of (7). (7) is in fact a clear expression of the thought which intuition rebels against, that that very set X could have had different members.

In §4 and §5 of this chapter, we will address the question of how the intuition of error in (7) and (5) is to be justified. For the moment, let us simply *use* the intuition to provide us with candidates for essential properties of sets: we will say that for each member y of a given set it is essential to that set that y belong to it. According to this principle, the membership of a set is the same at every world in which the set exists, so we can call the principle 'Membership Rigidity', or 'MR' for short. We formalize it as an instance of (S′), omitting the category predicate 'set' but using set variables:

(MR) $\Box(\forall X)\Box(\forall x)\Box(x \in X \rightarrow \Box(E(X) \rightarrow x \in X))$.

(MR) is inconsistent with the conception of sets as boxes with different members at different worlds; if '$\Box$' is read as 'always', it is also inconsistent with the conception of sets as boxes with different members at different times, a conception just as unintuitive as the corresponding modal one. Indeed, taking 'u' and 'w*' as terms for times, 'w*' standing for the present, the modal considerations which we have just worked through could all be reiterated for times, and this suggests that the explanation we give of our intuitions about the inappropriateness of the box metaphor should be able to explain why a set cannot change its members through time as well as why it cannot change them through worlds. We will see that this constraint on a successful explanation is satisfied below.

Although Membership Rigidity provides essential properties for sets, it does not yield individual essences. For a given set x, let us call the set M of properties decreed by (MR) to be essential to x, the set of 'membership properties' of x. That is, for each y in x, M contains the property of having y as a member; and to avoid complications with the subset relation, we will also include in M the property of having no members other than each of those objects y. Then (MR) does not by itself imply that for a given set x, the set M of membership properties of x satisfies both I(i) (that every member of M is essential to x) and I(ii) (that it is not possible that something else has every member of M). Obviously, M satisfies I(i), but the problem is with I(ii); that is, it is consistent with the truth of (MR) that the set M of membership properties of a set, even if it exhausts the non-trivial essential properties of that set, is nevertheless not an essence of that set. By inspection of I(ii), this must mean that (MR) is consistent with distinct sets having exactly the same members at certain worlds, and so we need some further principle to rule this out. The obvious candidate is (□E), which says that, in any world, sets with the same members are the same sets. But (□E) deals with only one case where distinct sets are alleged to have the same members, the case where the sets in question are both elements of the domain of the world at which they are claimed to have the same members. Yet it is also consistent with (MR), and with (MR) and (□E) together, that there be two sets X and Y such that the membership of X at any world where X exists is the same as the membership of Y at any world where Y exists, and there is *no* world where X and Y both exist; for instance, one who held the rather awkward view that individuals have transworld being but sets are world-bound individuals, would reject sameness of membership as sufficient for transworld identity, but could accept (MR) and (□E), since in virtue of the actualist treatment of quantifiers, 'X' and 'Y' in (□E) range over only the existents of one world at a time.

But from the nature of this counterexample to the claim that (MR) and (□E) provide essences for sets, it is clear how we should strengthen (□E) to obtain a principle which deals with it. We need a principle which says that if x is a set existing in a world u and y a set existing in a possibly distinct world v, then if x has the same members in u as y has in v, then x and y are the same set. According to this principle, if sets x and y have the same members, x in one world and y in another, then they are the same set. So we will label the principle 'Crossworld Extensionality', or 'CE' for short. Crossworld Extensionality does not have to be a component of every view which finds the box metaphor unsatisfactory, since both it and its negation are consistent with

Membership Rigidity; on the other hand, it does seem to be an element in the intuition that there is nothing more to the identity of a set than the identity of its members, which itself is inconsistent with the box metaphor. So we can regard the two principles (MR) and (CE) as distinct consequences of this more general view about the identity of sets, however it is to be elucidated and defended.

It is not wholly straightforward to formalize (CE) in modal language, even though we found it easy to state in possible worlds language above. For instance,

$$(8)\quad \Box(\forall X)(\forall Y)\,[(\Box(\forall z)(z \in X \leftrightarrow z \in Y)) \rightarrow X = Y]$$

does not have the effect of making a crossworld comparison between the memberships of two sets, since values of 'X' and 'Y' will always be chosen from a single world, that introduced by the evaluation of the initial '$\Box$', and compared at a single world, that introduced by the evaluation of the second '$\Box$'. Moreover, if we try to separate the initial pair of quantifiers with a '$\Box$', we obtain a principle which, though true, still does not provide the identity condition we want:

$$(9)\quad \Box(\forall X)\Box(\forall Y)[(\Box(\forall z)(z \in X \leftrightarrow z \in Y)) \rightarrow X = Y].$$

It is undeniable that if a set X which exists at u and a set Y which exists at v are such that, at *any* world, X and Y have the same members from amongst the existents of that world, then X and Y are the same set; for if X and Y are distinct and co-possible, then by Necessary Extensionality and the necessity of non-identities, this difference will manifest itself as a difference in membership within the domain of any world where they both exist; while if they are not co-possible, then at any world where one exists (the non-empty one, if the other is the empty set), none of the actuals of that world which belong to it will belong to the other (by the Falsehood Principle, since the other would not exist at that world). But in this argument for (9), we are envisaging worlds other than u and v at which at least one of X and Y exists, so we are implicitly *assuming* some way of picking out X and/or Y at such worlds, i.e., implicitly using Crossworld Extensionality, and it is the content of this assumption which we seek to formalize (this point also applies to (8), and to further strengthenings of (9)).

What we have to be able to do is to choose a value a for 'X' from one world u and a value b for 'Y' from another world v and then keep track of u and v somehow, so that we can speak of the members of a at u and the members of b at v. To do this, we can use the singly-indexed 'actually' operators introduced in §5 of Chapter 4, since the indexes enable us to keep track of the worlds where a and b have the memberships we wish to compare. The correct formalization is:

$$\text{(CE)} \ \Box_1(\forall X)\Box_2(\forall Y)([\Box(\forall z)((A_1(z \in X) \leftrightarrow A_2(z \in Y)))] \rightarrow X = Y).$$

The biconditional which is antecedent of the conditional in (CE) says that being in X in the first world is equivalent to being in Y in the second, so the conditional says that *if* this is so then X = Y, and the initial string of symbols makes this a sufficient condition for identity of any two possible sets X and Y and any pair of worlds such that X exists in the first world and Y in the second world of the pair. Thus (CE) identifies the sets X and Y which we imagined when we showed that (MR) and (□E) are not sufficient to provide sets with individual essences. Note that (CE) entails (□E) (consider the cases where the first and second worlds are the same), so one might regard (CE) as embodying the full strength of the component of the iterative conception which underpins (□E). Note also that the unindexed '□' in (CE) is essential, otherwise the variable 'z' will range over just the domain of the second world, and it may well be that the intersection of that domain with the membership of x in the first world is the same as the membership of y in the second, though they are different sets, since x has members at the first world which do not exist at the second.

In view of our definition of individual essence, we should expect to find principles analogous to (MR) and (□E) in the account of individual essence for objects of other categories. By I(i), we will look for a principle which yields non-trivial essential properties for entities of the relevant category, and then by I(ii) we may anticipate a principle like (CE) which rules out the possibility that distinct entities of the category agree on essential properties, a principle which specifies an indiscernibility condition whose holding across worlds is sufficient for transworld identity between the entities for which it holds. We will see that in fact we do find this pattern in the theory of individual essence for various categories of thing, but the indiscernibility principle is less intimately related to intraworld identity conditions than in the special case of sets, where (CE) entails (□E).

Finally, we should note an alternative way of adding to (MR) to obtain principles which fix essences for sets. In (MR) and (CE) we combine individually necessary conditions for transworld identity with a principle which makes them jointly sufficient, but we could achieve the same effect by replacing the sufficient condition for identity with a sufficient condition for existence; (CE) does not itself state an existence condition, since we are assuming, by our notational convention, that we are *given* sets as values of 'X' and 'Y'. But what might be a natural existence condition? If (MR) and (CE) are correct, then to do without (CE) we need an existence condition which implies it. A fairly natural transworld sufficient condition for existence which will imply (CE), given (MR) and (□E), is that a set exists in a world w if all its members at any given world where it exists, also exist at w. So if X has certain members at u, and these members all exist at w, then by the existence condition X exists at w, and by (MR) its members at w will be exactly the members it has at u. By (□E) no other set at w has exactly these members as well, so no counterexample to (CE) can arise. Note that in this argument, (MR) guarantees that it does not matter which X-containing world u we begin with.

The modal sentence which most directly expresses the sufficient condition for existence just formulated, a condition we call 'Set Existence', also uses the singly-indexed 'actually' operators in its formulation:

$$\text{(SE)}\quad \Box(\forall X)\Box_1[[\Diamond(E(X)\ \&\ (\forall y)(y \in X \rightarrow A_1(E(y))))] \rightarrow E(X)].$$

This formula says that for any possible X ('$\Box(\forall x)$') and any chosen world ('$\Box_1$'), if there is some world in which X exists such that every member of X in that world exists in the chosen world, then X exists in the chosen world. No indexed operator is needed to govern the consequent of the conditional, since in evaluating (SE), 'E(X)' would be evaluated with respect to the world introduced by the modal operator within whose immediate scope it lies, in this case the chosen world introduced by the indexed '□'. And what the argument one paragraph back shows is that (MR), (□E) and (SE) together imply (CE). However, (MR) and (CE) do not imply (SE) unless we assume a further principle. This is the principle that, given any objects at a world, all sets which can be built up from those objects also exist at that world; we might call this the principle of automatic set formation. If this principle in fact holds at every world, we can argue from (CE) to (SE) thus: if X's members at u also exist at v, then by automatic set formation at v, the set of them will exist at v, which by (CE) is the set X; so (SE) is verified, in fact without using (MR).

If we embed our modal theses about sets within a theory intended for structures with automatic set formation, there is little to choose between (SE) and (CE). However, automatic set formation is a controversial principle against the background of *de re* modality, at least, a more controversial principle than (CE), and has been challenged;[9] for someone might say that although the availability of the members of a set guarantees the possibility of forming the set, the formation operation need not actually be carried out. Thus a minimal *de re* theory of the essences of sets is better formulated with the less controversial (CE) as an axiom, since the truth of (MR) renders (SE) a conditional automatic formation principle: if X exists at w and its members at w exist at u, (SE) compels the formation at u of a set with exactly those members as *its* members, since it compels the existence of X at u and, by (MR), X must have the same membership at u as it does at w. There are also other reasons to prefer (CE) to (SE). For instance, if we use (SE), we will still need one of (CE)'s consequences, (□E), so an account appealing to (SE) is less economical. Secondly, for other categories of entity, an existence condition is less easy to state than a crossworld indiscernibility condition. And, thirdly, we will discover that the justification of such principles as these turns on features of the concept of identity, so that (SE) is less directly justified by the relevant considerations than a sufficient condition for identity like (CE); indeed, in view of the controversial aspect of (SE), it may not be justified by these considerations at all.

3. *The system MST*[10]

Although it is not to the main point of this chapter, it is of some interest to consider how *de re* modal theses about sets could be added to a standard axiomatic theory such as ZFI to obtain a system of modal set theory. We now present one such system, called MST. The reader with no interest in, or knowledge of, axiomatic set theory can proceed without loss of continuity to the next section of this chapter.

The language of MST, L(MST), is a first-order modal language with three predicate symbols, the one-place symbols 'S' and 'I', for being a set and being an individual, and the two-place symbol '∈' of set membership. Where possible, we will exploit our convention about variables to avoid use of 'S' and 'I', so L(MST) also has three sorts of

[9] See Parsons [1977].

[10] I developed the main lines of MST after reading Fine's 'Postscript' in Prior and Fine [1976], and also Fine [1977], but independently of Fine (1981b). However, I am indebted to the last of these papers for clarification of a number of points, especially on the confusing topic of abstracts, and to Fine himself for extensive comments on an earlier version of this material.

variables, general variables, set variables and individual variables, as explained above (the abbreviating formulae are in Form[L(MST)] as well as those abbreviated). The theory has *de re* axioms dealing with transworld facts about sets, individuals, and set membership, together with *de dicto* axioms dealing with the intraworld theory of sets. Our intention is that the intraworld theory should just be ZFI.

The first three *de re* axioms are (F), (MR), and (CE).

(F) $\Box(\forall x)\Box(\forall X)\Box(x \in X \rightarrow E(x) \& E(X))$.

(MR) $\Box(\forall x)\Box(\forall X)\Box(x \in X \rightarrow \Box(E(X) \rightarrow x \in X))$.

(CE) $\Box_1(\forall X)\Box_2(\forall Y)([\Box(\forall z)(A_1(z \in X) \leftrightarrow A_2(z \in Y))] \rightarrow X = Y)$.

The next two *de re* axioms deal with 'I' and 'X'. We have to decide whether it is merely essential to a set that it is a set, or whether it is necessary, *mutatis mutandis* for individuals. In the light of our discussion in §2, being a set at a world requires being formed at that world, which is to be brought into existence at that world, and thus we should hold '$\Box(\forall x)\Box(Sx \rightarrow E(x))$'. But then if sets were necessarily sets, every set would be a necessary existent. We therefore adopt the following Axiom of Set Rigidity:

(SR) $\Box(\forall x)\Box(Sx \rightarrow \Box(E(x) \rightarrow Sx))$.

If we also held the Falsehood Principle for 'I', we could infer an analogue of (SR) for individuals. An alternative course is to leave this question open and to take the essentiality of individuality as an axiom:

(IR) $\Box(\forall x)\Box(Ix \rightarrow \Box(E(x) \rightarrow Ix))$.[11]

To continue MST, we want to add axioms to the effect that ZFI is true at every world. We therefore add to the above axioms the *de dicto* necessitations of Pairing, Null Set, Power Set, Foundation, and Infinity, but not the principle ($\Box$E) already described, since it follows from (CE). Thus the empty set, which we denote 'Λ', is a necessary existent, by $\Box$[Null Set] (which gives us some empty set at each world) together with (CE), which implies that there is exactly one such set at each

[11] The axioms to date are in a sense 'modal' as opposed to 'set-theoretic', and are by no means all the axioms of this sort one might want to include. I leave it for the reader to ponder the addition of '$\Box(\forall x)\Box(Sx \rightarrow E(x))$' ('existence is necessary for being a set'), the analogous claim about individuals, the axiom of exclusivity '$-\Diamond(\exists x)(Sx \& Ix)$', and the question of what these exclude which is admitted as things stand, and what the logical relationships amongst the various modal axioms are.

world and that the empty set of any one world is the same set as that of any other. In fact, this follows from □[Infinity], one *de dicto* instance of Comprehension (see below), and (CE), hence □[Null Set] is strictly redundant. We add to this list the axiom '□(The existing individuals, if any, form a set)', i.e.

(IFS) $\Box(\exists X)(\forall y)(y \in X \leftrightarrow Iy)$

since only if the individuals are finite in number is the existence of a set of them guaranteed by the other axioms.

However, there are alternative ways to define the remaining axioms, since in their non-modal versions the defining principles, Comprehension and Replacement, are schemata, so questions arise about what restrictions are appropriate on their modal instances. Here are the non-modal principles:

Comprehension: For each formula ϕ with free variables among $x, Z, w_1, \ldots, w_k$,

$$(\forall Z)(\forall w_1)\ldots(\forall w_k)(\exists Y)(\forall x)(x \in Y \leftrightarrow x \in Z \,\&\, \phi)$$

is an axiom.

Replacement: For each formula ϕ with free variables among $x, Z, y, w_1, \ldots w_k$,

$$(\forall Z)(\forall w_1)\ldots(\forall w_k)[(\forall x \in Z)(\exists !y)\phi \rightarrow (\exists Y)(\forall x \in Z)(\exists y \in Y)\phi]$$

is an axiom.

For each principle, we have to consider whether or not its MST version should restrict substitutions for ϕ to non-modal formulae of L(MST), and whether or not the values of the parameters and Z should all be chosen from the domain of a single world. If the schemata are to yield only *de dicto* axioms, we must impose both restrictions on each. The intuitive idea that, at each world, all sets which can be built out of individuals from the domain of that world should exist, provides some guidance here. For example, the idea underlying Comprehension is that given any set X, any subset Y of X definable from parameters should exist. 'definable' means 'definable by an expression of the language', and, since the language we are working with allows modal formulae, a restriction of ϕ in the schema to non-modal formula would open up the possibility of there being a modally definable set which clearly exists, but whose existence is not a consequence of the axioms. As an example of such a set, we may consider the subset of X whose members are the contingently existing individuals in X. In other words, if given the existence of X,

(10) $(\exists Z)(\forall i)(i \in Z \leftrightarrow i \in X \,\&\, \Diamond(-E(i)))$

is to follow from the axioms, modal ϕ must be allowed in Comprehension. And as Fine has pointed out, if

(11) $\Box(\exists X)(\forall i)(i \in X \leftrightarrow \Diamond - E(i))$

is to hold, similarly in Replacement.[12]

The issue of parameters, that is, the issue of whether to insert '$\Box$' between each '$(\forall w_i)$', is less easily resolved. If a subset of a given set at a world can *only* be defined with parameters including non-existents at that world, it is unclear that it belongs in the cumulative hierarchy of that world, since non-existents then enter essentially into its generation, while our conception of the hierarchy is that the existence of every set in it is determined just by that of the entities at lower levels. But to allow merely possible parameters is not necessarily to admit such a set, since a condition can pick out a set without corresponding to the process which generates the set. The question is whether the Comprehension principle will have false instances if merely possible parameters are allowed. Note that in formulating non-modal Comprehension above, we might have restricted the (values of the) parameters to Z, but instead we allow them to range over the universe of sets. So in the modal case we will allow them to range over the modal universe of sets:

$\Box$[*Comprehension*]: For each formula ϕ with free variables among $x, z, w_1, \ldots, w_k$,

$$\Box(\forall w_1) \ldots \Box(\forall w_k)\Box(\forall Z)(\exists Y)(\forall x)(x \in Y \leftrightarrow x \in Z \,\&\, \phi)$$

is an axiom.

What of Replacement? The idea behind this principle is that certain sets exist because they can be defined as the result of replacing the members of a given set, one for one or one for many, by the entities which are to make up the new set (it follows from this that the cardinality of the new set cannot exceed that of the given set), and it does not seem to matter to this intuitive idea if the definition of the replacement relation has merely possible parameters. But will theories of different

[12] Fine [1981b, p. 186].

strengths result, depending on how we decide this question for □[Replacement]? Concerning some modal set theories based on ZFI but without (F), Fine has conjectured that when merely possible parameters are excluded, the sentence

(12) $\Box(\forall X)\Box(\exists Y)(\forall x)(x \in X \leftrightarrow x \in Y)$

is not a consequence of the axioms. (12) of course holds in MST, but for this system there is an analogous problem, that of whether (13) holds:

(13) $\Box(\forall X)\Box(\exists Y)(\forall x)[(\Diamond(x \in X)) \leftrightarrow x \in Y]$.

(13) will not hold in MST with *de dicto* Replacement if (12) is not a theorem of Fine's systems with *de dicto* Replacement; therefore, since (13) seems true—it says that given any possible set X and any world w, the existents of w which belong to X in some world form a set Y at w—we shall adopt the stronger formulation as the Replacement principle for MST:

> □[*Replacement*]: For each formula ϕ with free variables among $x, Z, y, w_1, \ldots w_k$,
>
> $\Box(\forall w_1) \ldots \Box(\forall w_k)\Box(\forall Z)[(\forall x \in Z)(\exists! y)\phi \rightarrow$
> $(\exists Y)(\forall x \in Z)(\exists y \in Y)\phi]$
>
> is an axiom.

This completes the listing of the axioms of MST.

A *model* for MST is a quantified S5 model in which all axioms of MST are true. More exactly, it is a 5-tuple ⟨W, D, d, v, w*⟩ such that every axiom of MST is true at w*, where W, D, d, and w* are as in Chapter 2 while v is an assignment function defined for L(MST): v('I') and v('S') are subsets of D and v('∈') is a subset of D × D. Since L(MST) contains no individual constants, the evaluation of a sentence of L(MST) in a model M proceeds through the expanded language L′, which contains a name for each entity in the domain D of M. Mod(MST) is the class of all models of MST, and a theorem of MST is a sentence true in every model in Mod(MST).

A rather obvious theorem of MST is

(14) $\Box(\forall x)\Box(\forall X)\Box(x \in X \rightarrow \Box(E(X) \rightarrow E(x)))$.

For if 'a∈b' holds at w, and 'E(b)' holds at u then, by (MR), 'a∈b' holds at u and thus by (F), 'E(a)' holds at u. However, to establish *de re* theorems of MST involving other Boolean notions customary in set theory, one must define these notions carefully. For instance we define the subset relation thus:

$$\alpha \subset \beta =_{df} S\alpha \;\&\; S\beta \;\&\; E\alpha \;\&\; E\beta \;\&\; (\forall v)(v \in \alpha \rightarrow v \in \beta)$$

where 'α' and 'β' are metalinguistic variables ranging over general and set variables of L(MST). Each condition in addition to the standard one is required here; for example, without the third conjunct, every non-existent set at w would be counted a subset at w of any existent set at w. With this definition, it then follows that

$$(15) \quad \Box(\forall X)\Box_1 [E(X) \rightarrow \Box(\forall Y)(Y \subset X \rightarrow A_1(Y \subset X))].$$

The reader is invited to develop the *de re* portion of MST for himself. We address the question of whether the *de dicto* portion of MST is stronger than ZFI below.

In standard set theory, it is usual to augment the operators of the formal language with the abstraction operator, with which we can form certain terms, set abstracts, for denoting sets. To be precise, L(MST) is expanded by addition of the new operator, and in the syntax of the expanded language (L(MST) with abstracts), we deal with this operator as follows:

(Abstraction): if v is a general variable and ϕ a formula, $\{v : \phi\}$ is a term with each occurrence of v in ϕ bound.

In a standard theory, '$\{v : \phi\}$' may be read 'the set of all objects v such that ϕ', and the operator does not add to the expressive power of the language, since any sentence σ containing abstraction terms can be translated into some sentence σ' which is free of abstraction terms and which is provably equivalent in the theory to σ.[13]

However, in modal set theory, the treatment of set abstracts requires some care. First, the reading of abstracts suggested above is ambiguous, in that in a modal context 'all' may be either an actualist or a possibilist quantifier. In our version of S5, quantifiers are actualist, and so, consonant with this, we admit only actualist abstracts: we read the abstract as 'the set of all existent objects such that . . .'. Secondly, when abstracts are admitted (whether possibilist or actualist) we thereby admit non-rigid singular terms; for example, '$\{x : I(x)\}$' stands for

[13] See Chapters 1 and 2 of Quine [1963].

different sets (the domain of existing individuals) at different worlds, since individuals may be contingent existents. Similarly, if we had constants 'a' and 'b' in the language, then at a world u where a and b both exist, '{a, b}', which abbreviates '$\{x: x = a \vee x = b\}$', would denote the set whose sole members are a and b, but at a world v where only a exists, '{a, b}' denotes at v the same set as is denoted at v by '{a}', the set whose sole member is a; so '{a, b}' denotes different sets at different worlds. Thus 'Socrates $\in$ {Socrates}' is false at worlds where Socrates does not exist even in theories without the Falsehood Principle, for under the actualist reading of set abstracts, the truth of that sentence would require Socrates to be a member of the empty set at worlds where he does not exist; but since Socrates is not a member of the empty set at worlds where he does exist (Λ has no existent member at any world) and since there must be such a world (every x in D is in some d(w)) this is inconsistent with (MR).

However, as in the standard case, the actualist abstract does not increase the expressive power of the language; in fact, the following clauses state equivalences on the basis of which abstracts may be eliminated (granted '$-\Diamond(\exists x)(Sx \,\&\, Ix)$');

(i) $I\{x: \phi\} \leftrightarrow \bot$

(ii) $S\{x: \phi\} \leftrightarrow \top$

(iii) $y \in \{x: \phi\} \leftrightarrow E(y) \,\&\, E(\{x: \phi\}) \,\&\, \phi\,[y/x]$

(iv) $E(\{x: \phi\}) \leftrightarrow (\exists z)(Sz \,\&\, z = \{x: \phi\})$

(v) $t = \{y: \phi\} \leftrightarrow (\forall z)(z \in t \leftrightarrow \phi\,[z/y])$, for any term t;

(vi) $\{x: \phi\} \in y \leftrightarrow Sy \,\&\, (\exists z)(Sz \,\&\, z = \{x: \phi\} \,\&\, z \in y)$

Here '$\top$' is a tautology and '$\bot$' its negation.[14]

[14] Clause (ii) may seem strange: do we really want to agree that

$$S\{x: -(x \in x)\} \leftrightarrow (P \vee -P)?$$

The problem is that in a model of MST, the Russell abstract has no denotation at any world. But to apply the predicate 'S' to an abstract is not to commit oneself to the existence, or even possible existence, of a denotation for the abstract: we can *stipulate*, for technical convenience, that 'the Russell set is a set' is to be true, even though there is no actual or possible set answering to the definite description here. If one were to give a *semantics* for abstracts, treating them as terms analogous to definite descriptions, then on a Scott-style approach, the necessarily non-denoting abstracts would be assigned objects not in the domain D of a model of MST, and in the resulting extended model the extension of S would include such objects. This is in fact Fine's procedure (see Scott [1967] and Fine

It is helpful to distinguish two attitudes to abstracts. On the first attitude, the abstraction notation is simply an abbreviatory device, the abbreviated sentence being the one which results when the abstracts are eliminated from the given sentence in accordance with the clauses above. From this point of view, the syntactic formation clause we gave for the notation is redundant: an abstract is well-formed in a given sentence iff a well-formed sentence of L(MST) results when the abstract is eliminated from that sentence. The formation clause is relevant only on the second attitude to abstracts, where we regard their addition to L(MST) as resulting in a new language. On this approach, the semantic account of a model of L(MST) must also be extended, so that every sentence of the expanded language can be evaluated; in particular, an evaluation clause for terms formed by abstraction must be stated. The second view then faces the problem of non-denoting terms: the quantifier rules for L(MST) with abstracts must be stated so that we avoid making inferences which are incorrect because a given abstract does not denote, and the evaluation clause must explain how truth-values are to be assigned to sentences with non-denoting abstracts. Since on the first attitude to abstracts, evaluation and inference are really carried out only with respect to unabbreviated sentences, so that our account of quantified S5 in Chapter 2 already suffices for the development of MST, the first attitude is the one we adopt.[15]

The distinction between the two attitudes is of obvious relevance to non-modal set-theory; if abstracts are primitive in the language, then the quantifier rules must be those of free logic; for if not, we could infer that there is a set which does not exist, '$(\exists Y)-(\exists X)(X=Y)$', which is a classical contradiction, from the theorem '$-(\exists X)(X=\{z: Sz \ \&\ -(z \in z)\})$', proved in ZFI by Russell's argument (note that in ZFI, the set of all non-self-membered individuals is just the set of all individuals). However, since the quantifier rules of MST are already free, it may be thought that in fact the complications ensuing from the second attitude to abstracts have been exaggerated. But this is not so, since an analogous problem arises in MST in virtue of the phenomenon of 'impossible sets'. These are of two categories. First, certain abstracts may fail to pick out a set at any world in some particular model for the unexpanded language L(MST) which is a member of Mod(MST), because

[1981b, p. 190]). But other approaches are also possible, e.g. a Russellian one on which clause (ii) would be inappropriate.

[15] I am not claiming that the second attitude is mistaken or unworkable. See Scott [1967] and Bencivenga [1976] for its implementation.

of the way objects are distributed throughout the domains of the worlds of that model: thus 'the set of all possible objects x such that possibly x exists' denotes a set at a world in a model M iff for some w in W, $d(w) = D$. However, we have avoided such troublesome abstracts by the actualist reading of the abstraction notation. But secondly, there are abstracts which fail to pick out a set at any world in any model because it is a theorem of MST that necessarily no set satisfies the condition abstracted upon: '$\{x: Sx \,\&\, -(x \in x)\}$' and '$\{x: x = x\}$' are examples. So the problems for the second attitude to abstracts still arise in MST, since from the necessary non-existence of the Russell set we could infer in quantified S5 the possible existence of something possibly identical to a necessary non-existent; that is, terms for necessary non-existents in quantified S5 are analogous to terms for non-existents in classical logic.

This last remark raises broader issues about the relationship between ZFI and MST. In particular, we should ask if MST is a *conservative extension* of ZFI. MST is conservative over ZFI iff:

(Con) For any sentence σ of L(ZFI),

$$(*)\ \text{ZFI} \models \sigma \text{ iff MST} \models \sigma.$$

Now it may seem clear that (*) fails left to right, for the logic of ZFI is classical and therefore

(16) $(\forall x)Ix \rightarrow (\exists x)Ix$

is valid, hence a theorem of ZFI, but (16) is not valid in free logic. However, (16) is nevertheless a theorem of MST, since Λ exists at every world of every M in Mod(MST). More generally, any closed sentence σ with no individual constants is such that '$(\exists x)E(x) \rightarrow \sigma$' is valid in free logic iff σ is valid in classical logic, and since we have '$E(\Lambda)$' as a theorem of ZFI, a set theory just like ZFI but for having the quantifiers of free logic will have the same theorems as ZFI (assuming the first attitude to abstracts). However, if L(MST) and L(ZFI) are augmented by constants which do not always denote necessary existents in MST, then in the statement of (Con), 'ZFI' should be understood as 'free ZFI'.

The proof of (Con) is straightforward. In the light of our recent remarks, the left to right direction is immediate, since all axioms of ZFI are clearly theorems of MST, their necessitations being axioms of MST. For the right to left direction, we need only prove the equivalent claim

(**) If σ is true in a model of ZFI, it is true in a model of MST.

However, (**) is also immediate, since any model of ZFI is *ipso facto* a model of MST. More exactly, let $M = \langle A, v \rangle$ be a model of ZFI in which σ is true. Define $M' = \langle D, W, d, v, w^* \rangle$ in which $D = A$ (A as in M), W contains just one world, which is therefore w*, $d(w) = A$ and v is as in M. Obviously, σ is true in M′, so it remains only to check that M′ is in Mod(ZFI). But this is also obvious, for there is only one world in M, all possible objects exist at it, and all ZFI axioms are true at it; thus the *de re* axioms of MST are automatically true—for instance, (CE) is true since Extensionality is true. Hence MST is conservative over ZFI.[16]

To bring our discussion of MST to an end, note finally that the above argument also establishes that MST is consistent if ZFI is. And, of course, MST is incomplete, as ZFI is, on account of Godel's First Incompleteness Theorem.

4. *Justifying Membership Rigidity: two unsuccessful attempts.*

For the special case of sets, (MR) alone refutes Quine's claim that anything can be changed to anything by easy stages, and in conjunction with (CE), provides a complete non-trivial account of the transworld identity conditions of sets. Given any possible sets x and y, x existing in u and y existing in v, we can say that x and y are the same iff they have the same members in their respective worlds. Of course, the members of a set may themselves be sets, but this does not render our claim about x and y *trivial.* If some of the members of x and y are sets, one can apply the same identity condition over again, and so on, until we reach sets whose members (if any) are not sets. If we have reached the empty set, there is no need to go further, while if we have reached non-sets, we will then need to apply a different theory of individual essence to say when *they* are identical across worlds; but however that should go, the question about the transworld identity of the original sets x and y has been completely resolved. In particular, if one restricts the objectual quantifiers to the pure sets, then (MR) and (CE) constitute a complete vindication of *de re* modality, a vindication which is extended as theories of essence are provided for further categories of thing.

[16] Many other questions about MST and its alternatives will undoubtedly occur to the reader; more topics are covered in Fine [1981b] than there is space to include here.

However, it is one thing to write down some intuitively appealing principles which for a particular category of entity meet Quine's challenge, another to produce an argument to justify these principles. Quine was not denying that arbitrary stipulations could block the transformation of anything into anything by easy stages, so we have to show that (MR) and (CE) are better than arbitrary stipulations. We will begin by looking at two attempts to justify (MR), one essentially due to Richard Sharvy, the other to David Wiggins.[17] Since the role of (CE) has not been widely noticed or discussed in the literature, we will not examine other arguments about it here, but in view of both principles' apparent connection with the intuitive idea that a set is 'nothing but' the collection of its members, we might expect a defence of (MR) to employ resources sufficient for the defence of (CE) as well. This will be true at least of our own account.

In fact, Sharvy addresses (MR) in its temporal rather than its modal guise; he seeks to explain why a set cannot change its membership through time. But it is easy to rewrite his explanation in modal terms. The crux of his view is this:

> That is . . . the reason why no class can change its members: an apparent variable class would have to be identical (at any given time) to some class that does not change its members. So at that time, Leibniz's Law would be violated.[18]

In effect, then, this is a temporal version of the modal argument Kripke gives for the necessity of identity, which as we saw in §5 of Chapter 3, threatens to force the counterpart theorist to deny Leibniz's Law. Sharvy's idea is that someone who allows sets to change their members from time to time, or equivalently, from world to world, will also have to deny Leibniz's Law when temporal or modal properties are admitted: for of the alleged identical sets, one will have the property of possibly having different members (or in the future having different members) while the other will not. But, in fact, there is no such implication. Consider the following simple model. We let $W = \{u, v\}$ and $D = \{a, b, X, Y\}$, where X and Y are sets and a and b are non-sets; we depart slightly from our usual conventions at this point in that we do not regard our use of distinct letters for sets in specifying D as carrying the implication that the sets are distinct. Next, let $d(u) = \{X, a\}$ and $d(v) = \{Y, b\}$, and let the extension of the set-membership relation at u be $\{\langle a, X\rangle\}$ and at v be $\{\langle b, Y\rangle\}$; finally, let $u = w^*$. Suppose now

[17] See Sharvy [1968] and Wiggins [1980, pp. 112–17].
[18] Sharvy [1968, p. 311].

that it is claimed that X = Y, so that (MR) fails. Then there is no contradiction with Leibniz's Law, since there is no property which this model represents as being possessed by X but not by Y, or conversely. In particular, X has the property of possibly containing b, as does Y, and Y has the property of possibly containing a, as does X. From Sharvy's point of view, the trouble with the model is that it does not contain a set which does not change its members through worlds: no set {a} exists at v.[19] But such a set cannot be forced upon an objector to Membership Rigidity, so it does not follow from Leibniz's Law that his position is mistaken.

Wiggins has a more complex, but no more successful, defence of Membership Rigidity. His idea is that the truth of (MR) derives from the necessary conditions for singling out this or that sort of entity and the consequential limitations on the extent to which an entity which must be singled out in such and such a way can be envisaged other than it actually is. So Wiggins claims that since a given set must be singled out as the possessor of such and such members, the set thus singled out cannot be envisaged with different members. Thus Wiggins gives a semantical–psychological interpretation of the idea that there is nothing more to the identity of a set than the identity of its members: to single out, or think of, a set, is to single out, or think of, a thing with these members. But this falls a long way short of justifying (MR). Singling out, or thinking of, a thing, is an activity of sentient beings carried out at particular worlds, and while it may be true that singling out a particular set X at two worlds u and v involves singling out its members at u and singling out its members at v, this fact does not imply that its members at u and its members at v have to be the same. Wiggins holds that since we *have* to single out a set *via* its members, we cannot envisage a singled-out set having different members, but this, again, is a *non-sequitur*. The necessity which Wiggins appeals to lies in the singling out relation, and it is quite consistent with there being only one way of thinking of an object, or way of singling it out, that the

[19] This remark should be elaborated for those who followed the discussion of abstracts in §3. The point does not really turn on choosing a world where a does not exist and applying (F). Suppose we admit a to d(v). Then we can still say that '{a}' denotes at u the same set as '{b}' denotes at v (X and Y respectively, these being the same set, so that a normal specification of the model would have used the same letter for both) and, further, automatic set-formation fails to hold at v. This means that {a} does not exist at v (as we say from 'outside' the model) though a non-standard reading of the abstract might allow '{a}' to denote singleton-b at v. Of course, this is absurd; the point is that there is no contradiction with Leibniz's Law, so appeal to the Law does not explain the absurdity.

properties of the object which figure in this way are contingent properties of it.[20]

Wiggins also holds that someone who denies Membership Rigidity does not really have any grounds for assenting to (□E), the necessitation of extensionality. He writes:

> Suppose someone doubted the necessity of the membership relation. How could he combine the doubt with a reasoned affirmation of extensionality, or advance on behalf of extensionality such claims as 'a set is nothing more than a unity constituted by its members' (Richmond Thomason, *Symbolic Logic*, London, Collier MacMillan 1970, p. 284)? If there is no other way of identifying such a unity than via its constituents, then its identity is derivative from these . . . There is no sense then in the idea of a set {x, y} with actual members x, y, turning up in another possible world lacking x or y.[21]

Again, it seems that if there is no other way of identifying such a unity than *via* its members, the most Wiggins is entitled to claim is that the unity's identity in the world in which it is being identified is derivative from its constituents in that world, which is consistent with its having different constituents in different worlds. Furthermore, this passage apparently claims that (□E) is some kind of consequence of (MR). But this is not so: it is Crossworld Extensionality, not Membership Rigidity, which implies (□E), and these two *de re* theses are logically independent: distinct sets can have the same members, provided they do not co-exist at any world, even though every set has the same members at every world where it exists, while indiscernible sets may always be identical across worlds, even when some set changes its membership through worlds in which it exists. Hence, despite the efforts of Sharvy and Wiggins, we still lack a justification for our views about the individual essences of sets.

5. *The grounding of identities and non-identities*

Our fundamental intuition about sets is that there is nothing more to their identities than the identities of their members. We used this intuition earlier to motivate (E) and (□E), but there is no obvious reason why this intuition should not be thought of as applying to transworld identity itself. (MR) and (CE) would then be seen simply as technical articulations of this idea. This move suggests two questions

[20] See footnote 10 of Chapter 6, and §2 of Chapter 9, for further discussion of Wiggins' views.

[21] Wiggins [1980, p. 113].

which may be pressed against the sceptic about (MR) and (CE). If there is nothing to the transworld identity of a set over and above the transworld identities of its members, we can ask of an alleged counterexample to (MR) what *makes* the two sets in the example one and the same. In the terms of our earlier case, for instance, we can ask what could make the set X of passengers on BA 167/9.19.82 in the actual world the *same* set as the set W of passengers on BA 167/9.19.82 in u, and no substantive answer to this question could avoid departing from our intuition about the identity of sets. Similarly, if presented with an alleged counterexample to (CE), an example in which a set X has the same members at a world u as a set Y has at v, and X and Y are claimed to be different sets, we can ask *in virtue of what* the difference between them arises. In general, then, we are asking the sceptic about our two principles for an account of that in which the alleged transworld identities and differences in his counterexamples *consist*; we are asking for the *grounds* of the claimed numerical difference or numerical identity in the putative refutations of (MR) and (CE).

It is difficult not to believe that our assent to (MR) and (CE) is motivated by such considerations as these. But we have still to find some underlying justification for these considerations, since without such a justification it is open to the sceptic to reject our demand for a description of the grounds of his identity judgements, or else to claim that the identity between, say, X and W above, is grounded in the fact that the passengers in X are on the same flight as the passengers in W. To be in a position effectively to counter either response, we need two things:

(I) We need an argument for the more general view that numerical identities and differences in transworld identity must be grounded in *some* way, so that the sceptic cannot reject our demand for his grounds.

(II) We need a convincing demonstration that someone who denies (MR) and (CE) deprives himself of any way of grounding facts about identity; in conjunction with (I), this refutes the imagined sceptical move of appealing to sameness of flight in our example.

Below, we make a start on these tasks, presenting the basics of a position which will be developed as the same issues arise in connection with the analogues of (MR) and (CE) in the theories of the individual essences of entities of categories other than set.

Let us start with (I). It is difficult to find an irresistible argument for the principle that facts about identities and differences must be grounded in some way, but the thesis that it is part of the content of our concept of identity, whether transworld or transtemporal, that there are no ungrounded facts about such identities, can be supported by illustration from a wide variety of cases.

Case 1. Consider the supposition that things could have been exactly as they are except that the steel tower in Paris opposite the Palais du Chaillot is different from the one actually there. To make sense of this supposition, it is not permitted to imagine that the tower is made of different metal from the metal which actually constitutes it, or that it has a different design, or designer, or history. The *only* respect in which the imagined situation is to differ from the actual world is in the identity of the tower. The extent to which such a difference seems unintelligible is some measure of the plausibility of the view that transworld differences must be grounded: in Dummett's terminology, the example shows the strangeness of the idea that there can be 'bare' differences in transworld identity; rather, there must be something in which such differences consist.

Case 2. What holds for differences also holds for identities. An interesting illustration of the peculiarity of 'bare' or 'primitive' transworld identity is provided by the hypothesis that an actually untwinned human might have been one of a pair of identical twins, or might have had an identical twin. Some facts of life are relevant here. Identical twins are produced when the normal processes consequent upon the formation of the zygote (the cell which is formed by the fusion of the father's sperm with the mother's egg) break down in such a way that after the first mitotic division of the zygote, the resulting daughter cells separate and develop into distinct embryos. The important fact is that mitotic division is physically symmetric, since each daughter cell receives a copy of the chromosomes of the parent cell, and each copy is, so to speak, semi-original. A chromosome consists of a pair of intertwined strings of DNA, and each string is itself a sequence of molecules called nucleotides. Replication is effected by the intertwined strings unravelling from each other, each string then acting as a template for the construction from materials present in the cell of a sequence of nucleotides exactly like the one from which the template string has just unravelled. Then each template string resumes the double helix structure by intertwining itself with the new string whose construction it has just directed. Thus two chromosomes are obtained from one, each new chromosome contains half the matter of the original chromosome, and the new matter in such a pair of daughter chromosomes has the same source. The two chromosomes now proceed to opposite ends of the cell nucleus, and since replication has been occurring with all the chromosomes in the cell, division of the cell down the middle produces a pair of cells indistinguishable in genetic content, each containing half the genetic material of the original cell.

Suppose, then, that A is an actual but untwinned human being. Could A have been an identical twin? Granted that by 'identical twin' we mean a pair of individuals produced by the above process, an affirmative answer to this question amounts to the claim that there is a possible world in which the zygote of A in that world undergoes fission as described, followed by the non-standard separation giving rise to two individuals B and C; and, furthermore, A is identical to B or to C. But we ought to be very reluctant to say that there is such a world. Of course, it is possible that A's zygote divides in the required way, but from the account of mitosis just sketched, it is evident that there is nothing in such a situation to determine which of B and C is identical to A; nothing grounds one of the identities rather than the other. Perhaps the example is rather underdescribed. There will be some worlds in which B turns out to be rather more like A as he is in the actual world than does C, but there will also be worlds in which the converse is true. This claim might be rejected, on the grounds that in each world where such twins are produced, we should identify A with whichever of the twins is more similar to A as he actually is. But in some worlds the twins may tie; furthermore, this procedure will make it impossible for A to be one of a pair of twins the other of whom is the more like A as he actually is, which seems wrong. Hence someone who wishes to speak of transworld *identity* has to say that the fact about which of the twins at a given world is identical to A is a primitive, ungrounded fact, a bare truth about the transworld identity relationship which is entirely lacking in actual traces, be they as verification-transcendent as you please.[22] Again, the degree to which the supposed identity is hard to grasp is a measure of the plausibility of the view that real identities are grounded.

Case 3. Case 2 may be regarded as a modal version of a temporal case involving the splitting of an amoeba, where it is natural to say that when the division occurs, the original amoeba ceases to exist and

[22] For a reader with Cartesian inclinations, the example can be reformulated to concern the question of whether my body could have been that of one of a pair of identical twins.

two new ones come into existence. It would be very strange to hold that, in fact, the parent amoeba survives the splitting and only one new amoeba comes into existence, and the strangeness of this view clearly derives from the impossibility of citing features in virtue of which the parent amoeba is identical to one rather than the other of the amoebae which results from the splitting.

Case 4. Cases 2 and 3 are of course reminiscent of examples discussed by writers on personal identity. Suppose Oldman's brain has functionally equivalent hemispheres storing the same memories, realizing the same abilities and character traits, etc., and imagine that each hemisphere is transplanted into a new body giving rise to two individuals, Newman-1 and Newman-2. Without appeal to some such entity as the soul, can we credibly maintain that one of Newman-1 and Newman-2 is identical to Oldman, while the other is not? Again, there are no features which could ground the putative identity, since the same features are realized in the two cases. This has not prevented the view that Oldman *is* identical to one or other of the new men from being held, but this primitive personal identity, not grounded in any feature we would normally regard as relevant to questions of personal identity, is certainly extremely puzzling.[23]

The description of these cases is by no means the last word on the doctrine that for each instance of identity or failure of identity, there must be facts in virtue of which that instance obtains, for in the next chapter we will consider two cases which are apparently inconsistent with the doctrine, and in Chapter 7 we will deal with two more. Nevertheless, enough has been said to lend the doctrine some plausibility, so we end our discussion of sets by indicating what the defender of Membership Rigidity and Crossworld Extensionality can do about the second argument he needs, as characterized under (II) above. That is, suppose the sceptic agrees that there must be features in virtue of which the identity he postulates between X and W obtains, and cites the fact that these sets contain the passengers of the *same flight* in the two worlds. How can this response be shown to be inadequate?

The problem for the sceptic is to produce consistent judgements about more complex cases without attributing absurd essential properties to sets. We will see the full extent of this problem in the next chapter but, for the moment, let us complicate our example by imagining a third world v in which our six individuals all exist but no one travels anywhere on the 19th or the 21st. In this world, we may assume

[23] See Chisholm [1970] for a defence of this view.

that the sets {a, b, c} and {d, e, f} both exist, so let us call them M and N. What should we say about the transworld identity relationships among X, Y, W, Z, M, and N? A sceptic who rejects (MR) because he thinks that it misidentifies the essential properties of sets may hold that M and N are not identical to any of the other four, on the grounds that it is essential to X to have members which travel on BA 167 on the 19th, and to Y to have members which travel on the 21st. But such a sceptic simply confuses the set of travellers on that flight with the property of travelling on that flight; it might be incorrect to translate everything he says into what we regard as the truth by construing his word 'set' as our word 'property', and to say that he lacks the concept of set, for he may use the word as we do in non-modal discourse; but we can still regard him as conflating two distinct notions in his modal discourse.

On the other hand, if the sceptic does identify M and N with some of the others, regardless of which, then he is open to the objection that he has not really provided grounds for his judgement that $X = W$ and $Y = Z$. By making either of the two possible identifications, he admits that being made up of passengers on the same flight is not necessary for identity with X; and if it is not sufficient, then obviously no grounds have been provided. And it seems unlikely that the use of contingent features of sets to provide conditions sufficient for transworld identity will lead to a coherent theory of these conditions. At best, we could make identity judgements on a case by case basis employing the contextually most salient features of the sets in the cases. It is hard to see how this position differs from one which denies both that transworld identity is a coherent notion and that there is a fact of the matter about the truth-values of *de re* modal sentences; so it would be incorrect to speak here of a theory of that which grounds transworld identity for sets.

It is obvious that these brief comments on scepticism about (MR) need elaboration, and we have not touched on (CE) at all yet. But rather than work through the relevant considerations in detail here, only to have to repeat them again in connection with entities of other categories, we will postpone elaboration of these fundamental ideas until our discussion of those other entities.

6

The Necessity of Origin

1. *Kripke's thesis*

IN THE three lectures gathered together under the title 'Naming and Necessity', Kripke pursues at least two quite distinct topics. One concerns the proper account of the semantic relation of reference, while the other concerns metaphysical problems about essential properties of individuals. Because Kripke frames his own account of reference in modal terms, and uses examples involving possibilities to refute two rival accounts, it is not immediately obvious that his positions about these two issues are independent.[1] But this should be clear at the present stage of our discussion; for instance, we have already seen that the necessity of identity says that one thing could not have been many, nor many one, and has little to do with whether or not proper names are rigid designators.

Once we have a clear view of what the necessity of identity says, it appears to be a thesis not easy to resist; however, in the same discussion Kripke introduces another claim which is more properly termed essentialist,[2] and which is considerably more controversial:

> The question [is] . . . could the Queen—could this woman herself—have been born of different parents from the parents from whom she actually came? Could she, let's say, have been the daughter instead of Mr. and Mrs. Truman? . . . we can imagine discovering this . . . But let's suppose that such a discovery is not in fact the case. Let's suppose that the Queen really did come from these parents . . . The people whose body tissues are sources of the biological sperm and egg . . . Perhaps in some possible world Mr. and Mrs. Truman even had a child who became Queen of England and was even passed off as the child of other parents. This would still not be a situation in which *this very woman* whom we

[1] See Salmon [1981, *passim*].

[2] The necessity of identity is not really an essentialist thesis, at least if we follow Fine in identifying essentialism with the view that individuals may be *distinguished* by their necessary properties; *every* individual is necessarily identical with itself. See Fine [1978b, pp. 288–9].

call 'Elizabeth II' was the child of Mr. and Mrs. Truman, or so it seems to me.[3]

It is probably these remarks, more than any other, which have brought about the renewed interest in essentialism in recent philosophy (the title usually attached to Kripke's doctrine here, 'the necessity of origin', is a small misnomer on the strong interpretation of 'necessarily', if the Falsehood Principle is applied to 'is a child of', but 'the essentiality of origin' is a more awkward phrase). If we generalize what Kripke says about the Queen, then he is arguing that the parents of any organism are essentially the parents of that organism. However, the identity of the parents of an organism, he says, is fixed by the identities of the bodies from which the sperm and egg come which gives rise to the organism; hence it is no counterexample to his claim that a sperm-and-egg transplant would result in an organism having different parents, in one sense of 'parent'. But in that case it is really the sperm and egg which matter: what is essential to the Queen is to come from the sperm and egg from which she actually came.

One cannot immediately extend this claim to every organism, since not all organisms are created by sexual reproduction. But they do all 'come from' some organic antecedent, which may be a single thing, such as the acorn from which an oak tree develops. For a general term to cover such antecedent entities, let us use the word 'propagule'; the oak tree's propagule is its acorn, while a human's propagule is his zygote, whose propagules are in turn the sperm and egg whose fusion that zygote is. Thus the relation 'x is a propagule of y', or 'Prop(x, y)' for short, can hold across either fission or fusion; we shall regard it as irreflexive, asymmetric and intransitive; and it is evidently a relation to which the Falsehood Principle applies, since an existent propagule cannot give rise to a non-existent, an existent cannot have a non-existent propagule, and two non-existents at a world cannot enter into the biochemical reactions of development at that world which make one thing a propagule of another there. Using this relation, we can formulate a general version of Kripke's views about the Queen as an instance of the essentialist schema (S′) (see §1 of Chapter 5) in the following way:

$$\text{(K)} \quad \Box(\forall x)\Box(\forall y)\Box(\mathrm{Prop}(x, y) \rightarrow \Box(E(y) \rightarrow \mathrm{Prop}(x, y))).$$

The reader should compare (K) with Membership Rigidity, and note

[3] Kripke [1972, pp. 312–14].

that, since we are not yet proposing any analogue to Crossworld Extensionality, the essentialism embodied in (K) does not attribute individual essences to organisms: it is consistent with (K) that at some world some organism has exactly the propagules which some distinct organism has at some other world. In addition, it should be emphasized that the question whether or not (K) is true is quite independent of one's views about the nature of the self: it would be beside the point to dispute Kripke's claims about the Queen on the grounds that the Cartesian self who is the Queen could have inhabited any old body. Rather, the Cartesian and others of that ilk, should read Kripke's remarks as claims about the Queen's body, albeit infelicitously expressed.

In what follows, we will be concerned mainly with the justification of (K). As with (MR), we will argue, this time in some detail, that one who denies (K) must deny that facts about the transworld identity of organisms are always capable of being grounded, or else he must propound some other equally implausible conception of identity. Our arguments will apply, *mutatis mutandis*, to the defence of Membership Rigidity, and we will then turn to the question of finding some analogue of Crossworld Extensionality. In addition, rather than simply relying on the intuitive discomfort one feels with the examples of alleged ungrounded identities in §5 of the previous chapter, we will consider two cases which are harder to deal with from the point of view of someone who holds that identities must be grounded, and our treatment of these cases will support application of this principle about identity in defence of (K).

2. *An unsuccessful defence of (K)*

Kripke has not himself given a detailed argument for (K), but others have attempted to do so, so we will begin this discussion by considering one of the best known accounts, due to Colin McGinn.[4] McGinn's strategy is to assimilate the origin relation amongst organisms to the identity relation, so that the necessity of origin becomes a special case of the necessity of identity. We will argue that this assimilation is illegitimate.

The biological relations McGinn considers are those of *continuity* and *d-continuity*. Continuity is a temporal, transitive, one-one relation—a human being, for example, is continuous with his zygote. d-continuity is like continuity, except that it need not be one-one; for

[4] McGinn [1976].

instance, an organism x is d-continuous with any number of organisms if these have undergone fusion to produce something with which x is continuous. To establish (K), it suffices to show that d-continuity is rigid, which in this context means that if a pair of objects is in its extension at one world where both of its members exist then that pair is in its extension at every world where the second member exists ('rigid' really ought to mean that the extension is the same at every world—thus '=' is rigid—or that a sequence of objects is in the extension either at every world where all the members of the sequence exist or at none such, but neither of these is quite what we want here).

McGinn argues that d-continuity is rigid in the following two steps:

Step 1. Continuity is necessary and sufficient for identity amongst organisms: a human and his zygote are one and the same; hence, since identity is rigid, so is continuity.

Step 2. d-continuity is like continuity in all relevant respects; so d-continuity is rigid as well.

The premiss that continuity is necessary and sufficient for identity amongst organisms is obviously the crucial one, but it is not at all plausible. McGinn's reason for holding it is that he thinks it would be unmotivated to deny the identity of the zygote with the resulting adult since

> . . . adults are commonly identical with children, children with infants, infants with fetuses, and fetuses with zygotes. Any attempt to break the obvious biological continuity here would surely be arbitrary.[5]

However, there is a non-arbitrary reason for denying the identity which does not involve denying any continuities. We have seen that the zygote reproduces by copying its own genetic content and then by dividing itself symmetrically in two. We have also seen that it would be inconsistent with the repudiation of bare truths about identity through time to hold that the zygote is identical with one or other of the resulting daughter cells. Since it cannot be identical to both, it follows that the zygote ceases to exist upon the completion of replication. But then it contradicts Leibniz's Law to identify the zygote with the resulting adult (or even the resulting embryo), since adults (embryos) *outlast* their zygotes, indeed, do not even temporally overlap them.

This argument is not irresistible. It is a familiar point that the indiscernibility of identicals does not give us a criterion of identity

[5] McGinn [1976, p. 133].

which can be used to resolve hard cases, since it often happens that we must first ascertain whether a certain F is identical to a certain G before we can decide whether or not the F in question has a property the G has. Here we have a case in point. Someone who holds that the zygote and the adult are the same will say that, after replication, the zygote no longer exists in just the Pickwickian sense in which the child the adult once was no longer exists: the person who was the child still exists, as an adult now, and the person (or human being, if zygotes aren't persons) who was the zygote still exists, as an adult; just as it is *strictly* false to say that the child no longer exists (becoming an adult is not equivalent to dying!) so it is false to say that the zygote no longer exists. That is, a defender of McGinn's position could claim that 'zygote' is a phase sortal for humans, rather than an ultimate sortal for organisms: each human being has a zygotehood which precedes his infancy and childhood, and the zygote, infant, child, and human are all one and the same thing.

It would certainly be unusual to use 'zygote' as a phase sortal, since in the mouths of biologists it is a predicate for a certain natural kind of cell, and it is an odd feature of McGinn's argument that it fails unless we stick to this idiosyncratic sense. But is there anything *wrong* with using 'zygote' as a phase sortal? There is certainly a disanalogy between the notion of zygotehood, on the one hand, and childhood, on the other, since zygotehood ends with the division, and hence ceasing to exist, of a specific entity, while there is no similar phenomenon in the passing away of childhood. Still, there is no *a priori* reason why phases should not end in the clear-cut way the purported phase of zygotehood does.

The real problem with this line of defence of McGinn is that when we treat 'zygote' as a phase sortal to save the identity claim he makes, the argument no longer suffices to establish what it aimed to establish, that it is essential to a given human being to have developed from the very same cell as that from which he actually developed: briefly, the phrase 'the zygote' will no longer refer to the entity, the propagule cell, to which it should refer if the argument for (K) is to succeed. For with talk of a phase of a human being there is an attendant distinction between the human and whatever makes him up during that phase. During childhood, the human (or his body, if you prefer to read (K) as a thesis about bodies but not persons) is made up of certain cells, but, by Leibniz's Law, he is not identical to the sum of the cells which make him up at any moment or during any period of his childhood, since the human can, and usually does, outlast any such sum; this would be true even if, as is not the case, exactly the same cells make up the human at each moment of his childhood. So if a human has a zygotehood before his childhood then, by the same reasoning, he is not identical to the sum of the cells making him up during that phase, that is, he is not identical to the single cell in question (nothing turns on its being one cell, for the same would be true if humans originated from a two-cell entity resulting from an association of sperm and egg in which each continues to exist as a separate entity). More specifically, if 'zygote' is used as a phase sortal, it will be correct to say that the human is identical with the relevant zygote, but the zygote to which he is identical is not the same thing as the propagule cell, for the cell ceases to exist when it divides, while the zygote does not; this is the crucial point —the cell is now an entity which constitutes the zygote, rather than the zygote itself, and the zygote no more ceases to exist when zygotehood ends than does the child when he becomes an adult.

So on this defence of McGinn it will be true, and therefore necesary, that the human is identical with the zygote he was—the human could not have been a different zygote, nor a different child—but for all that McGinn has said, the human could have originated from a different *cell*, since a different cell could have made him up when he was a zygote, just as different cells could have made him up when he was a child. However, the thesis which McGinn is supposed to be arguing for is the thesis that it is essential to the human to have originated from the very same propagule, i.e., the very same cell. So our conclusion is this: if we use 'zygote' as a sortal for cells, then it is false that a human is identical to his zygote; but if we use it as a phase sortal for humans (as we may legitimately do, if we wish), then although it is true that a human is (necessarily) identical with the zygote he was, this is insufficient to establish the modal relationship to the propagule itself of which (K) speaks. Either way, McGinn's argument fails to support (K).

Finally, a more general objection to McGinn's justification of (K) is that it does not extend in any obvious way to explain essentialist principles which are intuitively of the same kind as (K). For instance, we cannot justify Membership Rigidity on the grounds that sets are somehow identical to their members (which one?). Since any acceptable account of (K) should employ the same resources as are required for the explanation of true principles analogous to (K) about entities of other categories, we should try to develop a defence of (K) which satisfies this condition.

3. *The case of the moveable oak tree*

The simplest way to bring out the principles underlying (K) is to investigate the consequences of denying that origin is essential. So let us consider the case of an oak tree, for instance the oak tree which stands in the cloisters of New College, Oxford, and the acorn c which is that oak tree's propagule. In what follows, we use 'the cloisters of New College, Oxford' as a rigid designator of a place. The most favoured case for a sceptic about (K) seems to be the following. Imagine a possible world w in which there is an oak tree which grows in New College cloisters and which resembles the actual oak tree as far as is possible compatible with the supposition that this tree in w grew from an acorn c' distinct from c. Thus the c'-tree in w very quickly comes to be constituted of the same matter as the tree in the actual world, has the same morphology, etc. Suppose also that in w, c does not grow into any tree at all, or better, does not even exist. In sum, in the actual world there is a c-tree, in w a c'-tree, and the trees are indiscernible across these two worlds except only with respect to origin. Then, according to the sceptic about (K), it is sheer dogmatism to insist that these trees are numerically distinct. Can the identity of the propagule acorn really have that much significance?

We shall show that it does. Let us agree on some straightforward and uncontroversial possibilities for the acorn c, the propagule of the tree in the cloisters in the actual world. Let us agree that c could have been planted on the other side of the cloisters and could have developed into an oak tree there. We make no assumption about the identity of such a tree, nor about how much or how little any such tree would resemble the c-tree in the actual world. So consider the class of worlds where this happens. Some of these will be otherwise almost indistinguishable from w, since the planting of the acorn c on one side of the cloisters does not render it impossible that c' is planted on the other side and grows into an oak tree exactly like the c'-tree in w. Choose one of these two-tree worlds, u, say. Thus in u, as in w, the c'-tree is the one which bears a high degree of resemblance to the c-tree in the actual world, while in u the c-tree need not bear much resemblance to the actual tree beyond having its propagule. If we link trees which have a very high degree of resemblance in almost every respect by a continuous line, and label trees by their acorns, we have this picture:

c	c'	c, c'
w* .	w .	u .

We may now press the sceptic about (K) to answer the following question: which, if any, of the two trees in u is identical to the actual tree?[6] This question has only three possible answers consistent with the formal properties of identity:

(i) the c-tree in u is the actual tree;
(ii) the c′-tree in u is the actual tree;
(iii) neither of these trees in u is the actual tree.

We will now argue that whichever answer the sceptic returns, his earlier claim that the c′-tree in w is the actual tree commits him to the existence of bare facts about transworld identity, or to some even less plausible view if he tries to qualify his position.

Suppose the sceptic returns answer (i), that the c-tree in u is the actual tree. Then, since the trees in u are distinct, the c′-tree in u is not the actual tree, and so the c′-tree in u is not the same tree as the c′-tree in w, which, according to the sceptic, *is* the actual tree. But there is no difference whatever between these c′-trees; they have the same propagule, and by choice of u, they have the same shape, the same location, the same matter, and so on. Hence the sceptic must posit a transworld numerical difference where there is nothing in virtue of which this difference obtains. His position is exactly like that of the man who holds that a set X existing at one world can have exactly the members there that a set Y has at another world and yet not be the same set as Y. On each position, a very plausible sufficient condition for transworld identity, being indistinguishable in every 'intrinsic' respect, is being contravened. For sets, this principle is just (CE), since the only intrinsic feature of a set is its membership, but for trees, the identity condition is actually somewhat weaker than the analogue of (CE) which we will endorse for organisms, since organisms have more intrinsic features than those which will figure in this principle. So, in effect, the sceptic about (K) is in a worse position than is someone who just rejects a certain account of the individual essences of organisms, since the identity principle he is in conflict with here is not so controversial as the yet-to-be-formulated principle analogous to (CE), which will give a stronger sufficient condition for transworld identity.

Under this pressure, the sceptic may retreat to answers (ii) or (iii), on which, indeed, no ungrounded identities or non-identities will arise in connection with the three worlds just considered. But we can show

[6] *Contra* J. L. Mackie, we do not have to make any assumption about which of the two trees is the *better* claimant. See Mackie [1974, p. 560].

that under very plausible assumptions, these three worlds may be supplemented with a fourth, whose existence the *anti*-essentialist cannot consistently contest, and with this fourth world in the picture, ungrounded facts about identity reappear. The plausible assumption we need is that the c-tree in the actual world could have been just as the c-tree in u is, that is, it could have grown where the c-tree in u grows, and could have had that shape, matter, etc. Although this assumption is hardly controversial, in §4 below we will discuss the position of a certain type of essentialist who refuses to grant it (thereby making it essential to the actual tree to be different in some way from the c-tree of u). For the moment, suppose we do grant it, and from the class of worlds where the actual tree is just like the c-tree in u, let us choose a world v which differs from u as little as possible compatible with there being no c′-tree in it, and no other tree in the cloisters. Using the same conventions as before, we may extend the diagram above to the following:

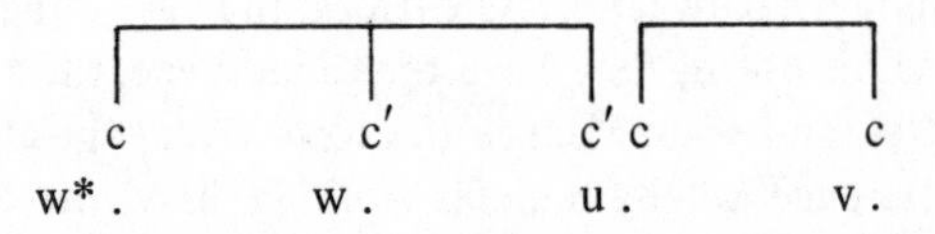

The c-trees of u and v are joined by a continuous line and are even *completely* indistinguishable in all intrinsic respects, by choice of v. But, according to the sceptic, they are distinct trees, since by hypothesis the c-tree in v *is* the actual tree, while on either answer (ii) or answer (iii) to our original question about u, the c-tree in that world is not the actual tree. So, once more, the sceptic must posit a transworld non-identity where there is nothing which grounds the difference, if he concedes that there are such worlds as w, u, and v. At the very least, then, we can defend (K) by saying that with a few other assumptions (K) is entailed by our intuition that identity must be grounded, the intuition we manifest if we agree that there is something wrong with the identity claims in the examples of §4 of Chapter 5.

4. *Intrinsic and extrinsic grounding*

The argument in §3 is the central component in the defence of (K), but it used one unexplained notion, that of an 'intrinsic' feature of an entity. The need for some such concept can be brought out by noticing that there is a response to our argument which pays some attention to the principle that identities must be grounded, yet which allows (K) to be rejected. Someone could say that there *is* something in virtue of which the c′-trees in w and u are distinguishable, and something in virtue of which the c-trees in u and v are distinguishable, and hence something in virtue of which these are pairs of numerically distinct trees, for the c′-tree in u shares the cloisters with a c-tree while the c′-tree in w does not, and the c-tree in u shares the cloisters with a c′-tree while the c-tree in v does not. We wish to say that these are not 'relevant' differentiating features because they are not intrinsic; but is this justifiable?

To hold that these relational differences between the trees are relevant to questions of identity is certainly unintuitive. How can what goes on concerning *another* acorn affect the identity of *this* tree? How can it be, for example, that if certain things had happened which actually didn't then a certain object would have existed, but if some other *causally isolated* process had also occurred, everything else remaining the same (so far as is possible), no such object would have existed? In natural terminology, we can say that the dispute here concerns whether transworld identity is only *intrinsically* or also *extrinsically* determinable. Consider someone who holds that even if certain features of objects and events are causally isolated at w from a given object x existing at w, these features may still be relevant to questions of the transworld identity of x with objects at other worlds; we say that such a person holds that transworld identity is extrinsically determinable, or *extrinsic*, for short. In fact, the thesis that identity is intrinsic, or indeed the thesis that it is extrinsic, are qualifications of the thesis that it is grounded, in the sense that they impose constraints upon the conditions which may be regarded as grounding an identity or non-identity. So to say that identity is intrinsic is to say that whether or not the identity relation holds across worlds between x and y must be settled by intrinsic features of x and y at the relevant worlds (Quine's criteria of continuity of displacement, distortion and chemical change would be examples of intrinsic grounds for transtemporal identity).

We can convince ourselves that our concept of identity does not permit its applications to be extrinsically grounded not only by testing our intuitions against examples, but also by contrasting identity with relations which are explicitly extrinsically grounded. For instance, if the counterpart y at a world w of some object x were selected by the criterion of maximizing similarity, it would be an extrinsic relation, since whether or not a given object at w is a counterpart of x would turn on whether or not *other* objects at w are more similar to x than it is. It is the extrinsicness of this relation, ultimately, which explains why a counterpart-theoretic semantics based on it ascribes the wrong truth-values to certain modal judgements, as illustrated in §5 of Chapter 3.

The intuition that identity relations are not extrinsic is especially strong in the case of identity through time. Suppose that throughout a period of time we are continuously observing a scene in which, it seems to us, there is a continuously existing object of some sort F which is undergoing no perceptible change. If identity through time were extrinsic, this sensory information would not even be *prima facie* evidence that the F observed at the beginning of the period was the same as the one observed at the end: we would also have to know how things were at the end of the period with causally unrelated F's existing then.

There is a certain kind of alleged counterexample to these claims about transtemporal identity. Suppose the ownership of a certain church building changes hands and what was an Episcopalian chapel becomes a Buddhist monastery. The signing of the documents, etc., which constitutes the change in ownership, may be regarded as a process causally isolated from the building (suppose it takes place on another continent). Nevertheless, before the signing we have one church, and after it, another; so whether the same church exists throughout the period of time in question may be determined by causally isolated factors.

However, there is a fallacy in this argument, of the same kind as one of those diagnosed by Wiggins in an argument for the thesis that identity is relative.[7] When we ask whether or not the same church exists throughout the period of time, we must decide whether 'same church' means 'same building' or, more strongly, 'home of the same denomination'. If in the above story we mean 'same building', there is no counter-example to the intrinsicness of identity, since the identity of the *building* (fixed by its location, design and the materials it is built from, say) does not alter when its ownership changes. And if we mean 'home of the same denomination' there is still no counterexample, since the issue of who owns the building is not at all causally isolated from the issue of which denomination it houses.

The intrinsicness of transtemporal identity, like its property of being grounded, may also be reinforced by an example about personal identity. Consider again the case of the split brain operation, and suppose

[7] See Wiggins [1980, pp. 30–5].

that it is actually performed and gives rise to Newman-1 and Newman-2 from Oldman. The operation might have been performed differently, just to the extent that the half-brain which now sits in Newman-2's head could have been thrown away instead, so that only one person would have resulted, with the half brain now in Newman-1's head. Suppose someone agrees with us that neither of the new men is Oldman, for the reason that identity must be grounded, but says that if only one half brain had been transplanted, the one in Newman-1's head, the resulting individual would have been Oldman. Then the actual Newman-1 is entitled to think, 'Thank goodness that other half-brain wasn't thrown away, otherwise I wouldn't have existed'. But in this thought the non-Cartesian can only grope for the reference of 'I', the entity which would not have existed if Newman-2 had not; for the person who would have existed would have been exactly as Newman-1 actually is in every physical and psychological respect.[8]

If we are willing to reject any view which commits its holder to the possibility of an extrinsic grounding for a fact about identity or non-identity, we are in a position to fill a *lacuna* in the argument of §3. It will be recalled that the sceptic who denied that the c-tree in u is the actual tree was shown to embrace bare truths about transworld identity by an argument which involved a world v at which the actual

[8] In Chapter 1 of Nozick [1981], the author advances at great length what he calls the 'closest continuer' theory of personal identity, which falls prey to this objection immediately, for if only Newman-1 results from the operation, he is the closest continuer of Oldman, while if both Newmen result, there is no unique closest continuer. Rather than regard the objection as damaging to his theory, however, Nozick prefers to say that we are here uncovering an antinomy in our notion of personal identity (it is the concept, rather than the philosopher's theory of it, which is at fault!). He admits (p. 47) that this must seem very *ad hoc*, and tries to dispel this reaction by drawing parallels between personal identity and other concepts for which there are both intrinsic and extrinsic analyses, where the latter appear superior. For example, Nozick holds that whether a belief is knowledge depends not just on the reliability of the methods by which the believer acquires the belief, but also on there being no non-reliable method whose role in the acquisition of the belief outweighs that of the reliable ones; so certain methods of belief-acquisition need not be sufficient for knowledge in a given case, even if in some other cases they do suffice, since those other cases satisfy the extrinsic condition that no additional method of a certain sort played such-and-such a role. But this comparison accomplishes nothing if personal identity just *is* a concept which requires an intrinisic account, as our intuitive reaction to the example in the text seems to imply. It is a significant difference between knowledge and personal identity that there is no analogous difficulty with an extrinsic account of knowledge: there is nothing problematic in saying that although Jones's belief that p is knowledge, if certain superstitions of his had played the main role in acquiring that belief, it would not have been knowledge. So the charge against Nozick that his antinomy claim is *ad hoc* still stands.

tree originates from its actual propagule at a place different from its place of origination in the actual world. Since a sceptic about (K) need not be an anti-essentialist, the defence of (K) could be blocked here by the doctrine that the place of origination of an organism is essential to it. Of course, this is highly counterintuitive, but the relevant question is whether a location essentialist ('L-essentialist') is making a *mistake*, or whether he is merely adopting some acceptable convention as an alternative to the one we happen to employ.

We find L-essentialism counterintuitive because of the lack of any very intimate connection between an organism and the place at which it originates; intuitively, someone who fixes transworld heirlines in terms of the locations of objects at worlds seems also to impute extrinsicness to transworld identity. But is there any connection between this sort of extrinsicness and the kind discussed above? There does seem to be a connection, a single underlying phenomenon, for we imagined the sceptic appealing to the properties of causally isolated entities to ground his identity claims, and the L-essentialist does the same. Certainly, the objects and features located at the place of an organism's conception need not be causally isolated from it, but these objects and features are quite distinct from the place itself. The place itself does not enter into any causal relation with the organism, because at least on the prescientific conception of place, places are of a nature such that necessarily they do not enter into any causal relations whatsoever. So L-essentialism is incompatible with the condition that identity be intrinsically determined.

There are other manoeuvres, of increasing complexity, which the sceptic about (K) may attempt. It is not very illuminating to pursue these, so one example will suffice. In our arguments against the sceptic, we have availed ourselves of worlds in which two propagules each give rise to organisms, and this suggests that there might be room for a qualified scepticism about (K) consistent with identity's being an intrinsically grounded relation. The sceptic could say that an organism can have one set of propagules in one world and a different set in another, if, but only if, there is some overlap between the two sets; then assuming the same propagule cannot function twice over, there will be no world in which each set independently gives rise to an organism. The simplest application of this qualified scepticism would be to creatures which reproduce sexually, such as humans, where there is an organism, the zygote, which has two propagules, the sperm and egg. Thus, if α is a zygote in w and β a zygote in u such that α's propagules in w are s and e while β's propagules in u are s$'$ and e, then the sceptic could identify α and β. The problem he faces is that if he can make this identification, then he should be able to identify β with the zygote γ in a world u, where γ's propagules in u are s$'$ and e$'$; but then, by his own principles, he cannot identify α and γ (contradicting the transitivity of identity) since there are worlds where both s and e and s$'$ and e$'$ fuse to form zygotes. It is perhaps possible to pick and choose one's identity judgements to avoid outright contradiction here, but hardly while remaining faithful to the principle that identity judgements must be grounded.[9]

5. *Essences and bare particulars*

These arguments establish that one who subscribes to the principle that the facts about transworld identities and non-identities must be intrinsically grounded will also have to subscribe to (K), on pain of commitment to rebarbative consequences. Moreover, similar points can be made about sets in connection with Membership Rigidity and Cross-world Extensionality. (CE) gives a sufficient condition for transworld identity between sets which grounds such identity in intrinsic features of sets (their membership), and since the effect of (MR) is to render the condition necessary as well, there can be no other intrinsic theory of the essences of sets which disagrees with this one, unless, as does not seem to be the case with sets, there is another family of intrinsic features which can be appealed to. More specifically, the reader who doubts the relevance of the arguments of §3 and §4 above to the case of sets will find that these arguments work effectively when instead of considering the trees which develop from the acorns c and c$'$, we consider the singleton sets whose sole members are, respectively, c and c$'$. If a sceptic about (MR) posits a transworld identity between these two singletons, then by introducing a world where they both exist in virtue of c and c$'$ both existing, we may press against him the question about the identities of the singletons in *that* world. His possible answers are the analogues of (i), (ii) and (iii) in §3, and analogous moves can be made against each answer. By intrinsicness, the identity of a set at a world cannot be sensitive to what other sets (outside the transitive closure of the given one) exist at that world, and if the set/attribute distinction is to be properly made, the identity of a set cannot be sensitive to properties of its members, or its members' members, etc.

[9] See sections 4–6 of my paper [1980] for further details.

Finally, a qualified scepticism which allows a many-membered set to change its members one at a time through a sequence of worlds succumbs to the argument of the previous section, since we eventually reach a world where by the transitivity of identity we still have the same set, but none of its members at that world are members of it at the world with which we started, giving rise to ungrounded identity again.[10]

The parallel with sets highlights the fact that we have still to complete our account of the individual *essences* of organisms, for until we have a principle analogous to (CE), it is left open that distinct organisms at different worlds have the same propagules. Mechanically transcribing (CE) into an analogous principle would yield a principle we might label 'Propagule Indiscernibility' (PI): if x at world u has the same propagules as y at world v then x = y. But this is incorrect. It is presumably true that more or less anything can develop into more or less anything, given sufficiently sophisticated engineering, so taking the acorn c which grows into a certain oak tree in the actual world, we can consider a world where c is treated in such a way that it develops into a small vegetable. Then (PI) entails that that oak tree could have been, e.g. a cabbage, and therefore that there are entities which can be oak trees in some world and cabbages in others. But we lack any conception of what such entities could be: they seem unattractively similar to the scholastic notion of 'bare' individuals, subjects of properties which can be abstracted from all the properties which 'inhere' in them. Bare individuals do not necessarily involve ungrounded transworld identity (see footnote 17), but are surely unintelligible in their own right. It would be possible to save (PI) by insisting that an acorn which grows into a small vegetable at a world *ipso facto* is not the same acorn as the actual one or, more strongly, that no propagule of a vegetable at one world can be a propagule of a tree at another. But it is unclear what could support this elimination of contingency in developmental outcome, since it may take only a very slight chemical change some time after the propagule has come into existence to produce the unnatural outcome.

The conception of a 'bare' individual arises in analogous temporal cases. Suppose a quantity of some polysaccharide is treated in such

[10] Ungrounded identity also features in the case where {a} and {b} are identified, for some a and b which are not co-possible. In this case, we cannot embarrass the sceptic with a world where both sets exist, granted that the existence of a set requires the existence of its members, but the sceptic will not be able to produce adequate grounds for the identity he posits.

a way that the sugar chains break down into their simple components; for instance, some cellulose decomposes into glucose. On one view, there is no transtemporal identity between the two quantities of substance in this example, but on another, there is: a single quantity of substance which is cellulose at one time becomes glucose at some later time. This second view appeals to the notion of an entity which can be different kinds of polysaccharide at different times, a notion which does not seem to answer to any concept we possess. Whatever the proper account of it is, there is a distinction we draw between changes which one and the same individual can undergo and changes which constitute the destruction of one individual and the creation of a new one; it may well be that this distinction is not completely defined, so that there are cases which its sense does not determine to be of one sort or of the other, but there are many more cases which the sense of the distinction settles: the zygote's mitotic division would be a case in point.

The limits on changes which are changes in a single individual are marked by what Wiggins has called 'substance' concepts[11] (or sortal concepts), and the fact that we do not have the idea of a bare individual manifests itself in the modal case as well as the temporal one, in the former by imposing limitations on what changes from world to world can be regarded as mapping out the contingent properties of a single individual; these limitations make the sort to which a thing belongs, or the kind of substance it is, essential to it. So we can modify (PI) to a principle of propagule-and-sort indiscernibility (PSI), which reads as follows:

If x is an organism at u with exactly the propagules $z_1 \ldots z_k$ and y is an organism at v with exactly the propagules $z_{k+1} \ldots z_{2k}$ then x

[11] See Wiggins [1980, Chapter 3]. I agree with much of what Wiggins has to say about identity through time, but am unsatisfied with his extension of his account to transworld identity. He holds that, given a sortal specification of what a thing is, we cannot conceive of *that* thing in a way which implies that it fails to satisfy the sortal, for this would be to conceive of it 'as having a different principle of individuation (different existence and persistence conditions) from its actual principle' (op. cit., p. 122). But why can we not 'just suppose' that the oak tree could have been a cabbage? We need a theory according to which our conception of the thisness of an individual is formed in the temporal case and then projected to transworld identity, to fix the boundaries of significance on *de re* hypotheses about the individual. Note that Wiggins obtains his essentialism about sets by counting the nature of a set's membership as a component of its principle of individuation; but it seems to me to be of a piece with this position about sets that the spatio-temporal route of a material object be counted essential to it. See § 2 of Chapter 9.

and y are the same organism iff (i) $z_i = z_{k+j}$, $1 \leqslant i \leqslant k$ for some $j \leqslant k$, and (ii) the sort of x at u is the same as the sort of y at v.

(PSI) is deliberately vague, in that it does not specify exactly how the sort of an organism is to be defined. In the time of a single world, the same individual can undergo a change of sex, but it is less clear that an individual of one sex could have been, from the outset, an individual of another (again, Cartesians may take the individuals here to be just the bodies). This appears to be the kind of boundary problem which our concepts are not sufficiently well-defined to settle, so the full story about individual essences of organisms is correspondingly left unfinished. But the *form* of the account is quite clear, and we may leave matters there, since it would serve no useful purpose to fix a boundary by stipulation.

6. *The branching conception of possible worlds*

Whatever account of an individual's essence we give, we rely to some extent on a parallel with identity through time for support for the thesis that the account is genuinely sufficient for transworld identity; for it will always be open to someone to say, concerning any non-trivial condition we argue to be sufficient, that he can conceive of *distinct* possible objects satisfying the condition. On this view, for instance, it would be held that distinct possible organisms can have exactly the same origin and be of exactly the same species and sex, one in one world and the other in another. The view in question is known as *Haecceitism*, since it attributes to each individual a primitive identity or thisness, as opposed to the kind of essentialism defended above, according to which non-trivial conditions sufficient for transworld identity can be given.

Some philosophers might say that Haecceitism is incorrect as far as transtemporal identity is concerned, for it seems possible to give criteria for transtemporal identity which are both necessary and sufficient. For instance, a continuity account, a version of which was adverted to by Quine, has some degree of plausibility. The classical conception of transtemporal identity as spatio-temporal continuity may be stated as follows:

(C) For any sortal F and any objects x and y, x and y are the same F iff x is an F and y is an F and for any times t, t′, if x exists at t and y at t′, then for each time t″ between t and t′ there is a region of space occupied by an F at t″ such that the interior of the sum of these regions (for all the t″) is a continuous region of space.[12]

A Haecceitist about identity through time would have to claim that even if p is a continuous path through space which is occupied at every instant of the period of tracing by an F, nevertheless distinct portions of p may be occupied by distinct F's. And it seems natural to challenge both the modal and the temporal Haecceitist in the way we challenged the sceptics about (MR) and (K); it is for them to explain in what the distinctions they draw consist.

Nevertheless, even if Haecceitism about identity through time is mistaken (we have still to see if it is), it may not be that Haecceitism about transworld identity is mistaken, and so far, for both transtemporal and transworld identity, we have only given some examples to prompt the intuition that a demand for grounds is always justified. It is time to consider some harder cases for our view about both kinds of identity. In this section, we will discuss an alleged example of ungrounded facts about the transworld relation, due to Robert Adams, and in the next, an apparent example of ungrounded facts about the transtemporal relation, due to Kripke.[13]

In Adams's example, we consider a world w in which there are two qualitatively indiscernible iron globes which have always and will always exist; that is all there is to w. But neither globe is essentially immortal, there are no restrictions on the times at which either globe could cease to exist, and the existence of either is in no way tied to the existence of the other. Thus there are worlds u and v just like w, except that in u one of the globes ceases to exist at a time t (before time ends, if it does) while in v it is the other globe which ceases to exist then (the assumption that the globes are indiscernibles in w is not essential, but simplifies the story).[14] In this set-up, according to Adams, the facts

[12] Note that in (C), the continuity condition has to hold for *every* interval [t, t′] such that x exists at t and y at t′. Note also that (C) nowhere quantifies over 'time-slices' or 'instant-stages' of ordinary continuants. Rather, it embodies Wiggins's conception of transtemporal identity as spatio-temporal coincidence under a concept.

[13] The Adams example is from Adams [1979, pp. 22–3]. For Kripke's example, I rely on Shoemaker's account in Shoemaker [1979, pp. 327–8].

[14] To sidestep the issue of whether or not the intraworld Identity of Indiscernibles is true, the example may be changed in the following way: in the 2-globe world w, let the globes be differentiated by contingent properties. Then even if it is necessary that if the globes co-exist then they are differentiated by some property, it is consistent to postulate worlds u and v, u with one globe and v

about transworld identity are primitive, i.e. ungrounded, because any feature we might appeal to as sufficing for the identity of the globe in u with one of the globes in w also holds between the u-globe and the other globe in w; *mutatis mutandis* for the v-globe. And the numerical difference between the u-globe and the v-globe is consequently also ungrounded, for no intrinsic feature differentiates either of these globes in its world from the other in its.

But these conclusions are unwarranted, and are at odds with the natural way of thinking about the globes, on which we can explain the facts about transworld identity in terms of identity through time. That is, one thinks of w as a course of events and of u and v as courses of events 'branching' from w at the time t when one globe ceases to exist in u and the other in v. Thus the transworld identities are explained by transtemporal identities across the branch-point at t. Before t, the very same course of events constitutes w, u, and v, and if we trace back in u from some point after t into w, and trace back in v from some point after t into w, we arrive at different globes; so the transworld difference between the u-globe and v-globe is explained by the intraworld numerical difference of the globes in w together with the branching conception of the worlds. This conception thus eliminates the appearance of ungroundedness in the facts about transworld identity in Adams's example.

We would like to generalize the branching conception of these three worlds to meet a certain objection to the theory of individual essence we have advanced. Although in giving the essence of an individual object we have not rendered the account trivial by appealing to the identity of that object itself, we have allowed non-qualitative properties to enter into essences; for instance, it is part of the essence of the actual oak tree in New College Cloisters to have originated from the acorn c, and no other. As a result of this, it may seem that we have not really shown that facts about transworld identity are grounded, since the transworld identity conditions of the objects which enter into the essences of other objects may themselves be ungrounded. But the branching conception provides an assurance, for a wide range of categories of object, that this is not so, since, provided the objects which enter into the essences of other objects in some sense themselves 'come

with the other, such that the postulated changes in contingent properties in the u-globe from w to u, and in the v-globe from w to v, yield the required crossworld indiscernibility of globes between u and v. In this situation, the interpretation of our understanding of the distinction between u and v advanced below is still applicable.

from' yet other objects, in a way which eventually leads to a temporal regress, we will at some stage in this regress be able to explain all relevant transworld identities as transtemporal identities holding across a branch point, just as the identities in Adams's example were explained.

The generalization of the branching conception we want is this: if u and v are worlds which at any time have some existent object in common, then u and v have some initial segment of their courses of history in common. In the light of this thesis, we see that the function of pairs of principles such as (K) and (PSI) is to enable all facts about transworld identity between u and v for objects which come into existence after the branch point, to be completely fixed by the content of the initial segment which u and v have in common (which may extend infinitely backwards in time).[15]

We can use Adams's example to refine the branching conception further. For instance if the globes in w are contingent existents, then there is a world w′ which is just like w except that in it only one of the w-globes exists. It is then impossible for w and w′ to have an initial segment in common, since at any time there are two globes in w and only one in w′, and so it would follow, by the generalized branching conception, that there is no world in which only one of *those* globes exists, contradicting our initial specification that the globes are not necessarily sometimes co-existent. To deal with this difficulty, we need the notion of a *separable* course of events in the history of a world w, a notion which will enable us to count amongst the worlds branching from w, worlds which consist in or extend a separable course of events in w. Causal isolation would be one criterion of separability, and assuming that the w-globes do not causally interact, our world w′ will also be a world branching from w. But other alleged worlds will be excluded, even when we have the notion of separability. For instance, it might be claimed that there is a world u which is just like w except that the iron which constitutes the two globes in w constitutes three globes in u, globes which, like the w-globes, have always existed. Here we reach the limits of intelligibility the branching conception imposes, as we come up once more against an ungrounded transworld identity, that between the quantities of iron in w and in u. On the view which we are presently defending, there is no such world as u.

[15] The branching conception has been discussed in a number of places by Hintikka, although he would not agree with my view of its range of applicability. See, e.g., Hintikka [1975, Ch. 2]. My view raises the problem of transworld identity for times themselves, which I address briefly at pp. 84–5 of Forbes [1981].

We said above that pairs of essentialist principles function to allow the content of a common initial segment to fix transworld identities amongst later objects. Thus, any particular transworld identity will be grounded either by a transtemporal identity involving those objects themselves, as in Adams's case, or else by facts about ancestry and kind and transtemporal identities amongst entities at some earlier stage in an ancestral tree. So the essence of an object involves those other objects through which we make the first step in tracing back to resolve a question about transworld identity for that object. However, it is conceivable that an object lacks any such essence, for an object may be in a certain sense 'simple'. A simple object would be one which in no sense 'comes from' any other objects, and if we say in Adams's example that the globes come from the quantities of iron which constitute them, then those quantities would themselves be examples of simple objects, objects without individual essences.[16] But it would be a mistake to infer from this that facts about transworld identity for simple objects may be ungrounded. In Adams's example the facts about transworld identity for the quantities of iron were also fixed by transtemporal identities, and we rejected the coherence of the hypothesis of a world in which those quantities make up three globes rather than two, if it is stipulated that there have always been these three globes, as opposed to their having come about *via*, say, one of the w-globes dividing. Hence the theory of individual essence we are propounding applies only to categories of object whose members may be said to 'come from' other objects in some fairly natural sense, as is exemplified by biological development or set-theoretic containment. But our claim that identity is an intrinsically grounded relation is not restricted to objects of these categories, as is manifested by its applicability to simples.[17]

7. *A problem about identity through time*[18]

The branching conception of possible worlds allows some cases of transworld identity to be directly grounded in transtemporal identities, but

[16] We could appeal to the molecules or atoms of which those quantities of iron are composed, but unless atomism is *a priori* false, the conceptual problem of simple objects will eventually rearise.

[17] Thus someone who thinks that bare individuals can be abstracted from objects with properties need not contradict our thesis that identity is intrinsically grounded if he makes no *de re* judgements about such individuals which require for their truth that there be worlds in which the same bare individual exists only at times after these worlds diverge.

[18] In the section which follows, the reader should separate two questions in

we have done no more to defend the view that *this* type of identity is an intrinsically grounded relation than, again, to give some examples of alleged ungrounded transtemporal identities and to point to their peculiarity. This strategy would be easy to outflank if there were quite straightforward examples of ungrounded identity through time and, according to Kripke, there are indeed such examples. Consider a perfectly homogeneous sphere at a fixed location rotating with constant angular velocity through an interval of time $[t_1, t_2]$, and compare the following sequences of half-spheres. The first sequence s_1 consists of a half-sphere for each instant i in $[t_1, t_2]$, the half-sphere which at i occupies the region r occupied at t_1 by the eastern portion of the sphere. Since the sphere is rotating, no half-sphere will appear in this sequence more than once, unless the interval is long enough to allow the sphere to complete a revolution (assume not). The second sequence s_2 is the sequence of half-spheres which would have occupied the region r if the sphere had halted at t_1 and remained stationary through t_2, i.e. it is the constant sequence of a single half-sphere, each occurrence of this half-sphere in the sequence being associated with a slightly different region of space, the region occupied by the half-sphere at the instant i indexing the given occurrence of the half-sphere in the sequence.

Since the sphere is in fact rotating, the half-sphere in r at t_1 is distinct from the half-sphere in r at t_2, but this transtemporal difference appears to be ungrounded. To see why, suppose we try to use the spatio-temporal continuity analysis of transtemporal identity ((C) above) to explain the numerical difference of these two half-spheres. Then we find that they are actually *identified* by this account, since the sum of the regions occupied by the half-spheres at the associated instants i in the sequence s_1 is of course a continuous region: it is just the region r itself, and this region is continuously occupied by a half-sphere of the appropriate sort (fixed by the dimensions of r). At t_2, the half-sphere which was in r at t_1 is in some other region of space,

his mind. One is whether Kripke has given a counterexample to the continuity account of transtemporal identity, and the other is whether he has given an example of a bare identity of a sort some analogue of which could arise in the modal case (if we were ingenious enough to think of it) to cause trouble for the principle upon which our defence of essentialism has been premised. The arguments in the text are intended to justify answering the second question in the negative, though they are admittedly less conclusive with respect to the first, in this context, less important, question. I am grateful to Christopher Peacocke for criticism of an earlier version of this section; the suggestion about atomism *ad fin* is based on a speculation of his.

and, of course, tracing that half-sphere also yields a spatio-temporally continuous path, the path determined by the constant sequence s_2, i.e. the path traced through the interval by the half-sphere which was in r at t_1. But what the account of transtemporal identity in (C) fails to do is to give us a reason to count *this* path as the path of a single half-sphere rather than the path consisting in just the region r, which we know would be the path of a single half-sphere only if the sphere had been stationary.[19]

A possible reaction to this case is to look for further features to ground transtemporal identity, features which get the case right, and to use these features to strengthen our criterion for transtemporal identity. It seems that the question of which path is the path of a single half-sphere depends upon the angular velocity or, more generally, upon the motion properties, of the half-spheres in each sequence. For instance, if at every i in $[t_1, t_2]$, each half-sphere in a sequence constructed like s_1 is at rest, then we know that the region determined by that sequence (r itself) is indeed the path 'followed' by a single sphere during $[t_1, t_2]$; so perhaps we can add something to the continuity criterion which speaks of motion properties. But Kripke could fairly object that appeal to motion properties to *ground* facts about transtemporal identity is circular, since concepts of motion are defined in ways which require the *application* of transtemporal identity; the simplest example is that of the linear velocity of an object at a time, which is the limit of a sequence of average velocities, each average velocity being the distance travelled by *that* object *during* a certain interval, divided by the magnitude of that interval. Indeed, to speak of *two* groups of facts here, those about the transtemporal identity

[19] In Hirsch [1971] the author gives a set of rules for tracing the careers of objects through time under sortal concepts along a path P, one of which is the 'No Choice Requirement': there is no path P′ such that F is instanced on every point of P′, and P′ is spatio-temporally continuous, and P and P′ partly coincide and partly diverge (p. 36). But he wishes to allow that in cases where we do have a choice we may make one non-arbitrarily in accordance with the criterion of coherence with identity judgements by the other rules. However, Kripke's example does not require an extension of an already partly determined notion of identity: it is fundamental to that notion. So the No Choice Requirement could not be used to solve it. A similar remark applies to the suggestion that there is no fact of the matter about what the correct identity judgements are in the example. Another possible reaction is to query the genuineness of the sortals the example uses, 'half-sphere in region r at t′, etc., but this reaction also seems to me to lack credibility: we really have no difficulty in conceiving of the objects in terms of which the example is formulated, and in grasping the idea of a spatial route followed by such an object through an interval of time.

relationships amongst the half-spheres and those about the motion properties of the half-spheres, appears mistaken: there are just two different ways of formulating the same facts.[20]

Another possible reaction to the example is to say that our judgements of transtemporal identity are guided by how we think the half-sphere *would* behave, were the other half somehow taken away, yielding circumstances in which the continuity criterion by itself would give the right answer. But however this idea is worked up it seems to reverse the facts about what grounds what: we do not think that the truth of the counterfactual grounds the identity facts in the actual case; rather, the counterfactual about where that half would be is true *because* of the actual identity facts.

It would not be right to think that the example turns on some special feature of rotation, since the same problems arise if we consider a segment g of a homogeneous, rigid rod moving through space on a straight path, and choose an interval of time at each moment of which some segment or other of the rod occupies the region which was occupied by g at the beginning of the interval. But by comparing this case with Kripke's, certain common features emerge as suggestive. First, in each case the problematic objects are singled out by sortals which refer to regions of spacetime, a procedure imposed by the fact that the objects in question (half-spheres, rod segments) are not wholly circumscribed by physically proper boundaries; e.g., there is no physical mark of the distinction between one rod segment and another. It may be that such objects form a conceptually special category, for which there are non-*ad hoc* reasons to complicate the account of their transtemporal identity. Secondly, in the specifications of each example, we stipulate particular motion properties for some *other* object of which the problematic objects (half-spheres, rod segments) are *part*; and this suggests that if we may assume that the motion of the whole applies to the parts as well, we can derive consequences for the transtemporal identity relationships amongst the parts without circularity.

Unfortunately, the idea that the motion properties for wholes

[20] In Shoemaker [1979] the author tries to get round this point by defining motion concepts for spatio-temporally continuous sequences of instantaneous thing-stages. But to arrive at our genuine concepts of motion he has to be able to distinguish those sequences which make up continuants from those which do not, and to do this, as he himself recognizes, he has to rely on a notion of causal connection which is itself defined in a way which presumes upon the notion of transtemporal identity (see pp. 329–30 and 336–7). So the detour through sequences of thing-stages accomplishes nothing.

ground transtemporal identity for parts cannot be non-circularly implemented. Consider the case of the rod: from the fact that the rod has moved in such-and-such a way during a certain interval (the movement specified by the displacements of its end-points), nothing follows about which rod-segments singled out at one time are identical with which segments singled out at a later time, unless we know the facts about the relative spatial distances amongst the rod segments throughout the interval, facts which presuppose transtemporal identity for the rod segments. By specifying that the rod is *rigid*, of course, we fix what those facts are, but this specification just uses the notion we are trying to elucidate: it specifies, e.g., that the distance between this segment and that one is the same *throughout* the relevant period of motion.

Nevertheless, there is an urge to resist accepting that Kripke has given a case which refutes the thesis that identities and non-identities must be intrinsically grounded, since this would mean that the identities amongst half-spheres in his case are of the same sort as those in the examples of bare identities given earlier, for instance, the alleged identity between Oldman and one or other of the Newmen. But it seems clear that the true identity judgement about the half-spheres is capable of being supported in a way in which the judgements in our paradigm examples of bare identities are not: we simply do not find the former mysterious in the way we do the latter. What we should try to do, therefore, is to pin down exactly what the differentiating feature of the two cases amounts to, with a view to deriving some account of the intrinsic grounds of the identity facts in Kripke's case from that feature.

One who holds that the facts about the spatial routes traced by the parts through space during the relevant interval are themselves bare facts posits a certain analogy between the two cases, for he could describe the case of persons as one in which the same self is in some kind of quasi-motion through the space of bodies, and as a matter of bare fact, moves from Oldman to Newman 1. The reason why we are inclined to reject such an analogy, presumably, is that ascription of Oldman's identity to Newman 1 has no consequences of either an actual *or* a dispositional nature in which the correctness of that ascription, as opposed to the other one, could manifest itself; nor do we understand what difference in initial conditions might bring about migration to Newman 1 *rather than* to Newman 2. But in the motion cases, different claims about the transtemporal identity conditions of the parts equate with different motion properties, and *these* differences can certainly be expected to have at least dispositional manifestations, the exact nature of which will depend on the laws of nature; consider, for instance, how we would expect an object to behave in a Newtonian universe were it to collide with a segment of the sphere, one which is rotating with a given angular velocity, rather than stationary. We also understand how initial conditions could differ, in terms of forces acting on bodies (in this case, torques), so that in one case we eventually get a stationary sphere, and in the other, a rotating one; admittedly, this does not apply to the Newtonian possibility of two worlds differing actualistically only with respect to the value of the constant angular velocity of a certain sphere, the sole occupant of each world, but in this case, whose prescientific intelligibility might be doubted, there are still the dispositional differences.

Might such dispositional facts be appealed to as the grounds of the identity facts? Normally, one would reject such an appeal, for the reason that dispositional facts about an object must themselves have categorical grounds in the nature of the object or its current properties. But in the present example, where it is beginning to look as if there are no other candidates for grounds, there is a case to be made for grounding the identity facts in the dispositional ones. The undisputed datum is that we understand the distinction between a situation in which one account of transtemporal identity amongst the half-spheres is the right one and a situation in which another is the right one. Moreover, the terms in which this understanding is given are, *ipso facto*, those which will specify the required grounds. But the idea that there are no terms other than identity itself in which the understanding is given is difficult to comprehend, for these identity questions are not open to being settled just by observation, and this makes it quite mysterious how we could come to *have* such an understanding of the difference between the two situations: to say that there is nothing in which our understanding consists seems little different, in this instance, from saying that our understanding is empty of content. So when pressed as to what it means to say that the identity facts are these rather than those, we may well turn to dispositional differences as constitutive of the distinction. The details of such differences are of course *a posteriori*, but the claim would just be that what it is for the identity facts to be one way rather than another is for the facts about what would happen were certain interactions involving the relevant object to occur, to be one way rather than another, according to whatever the laws of nature happen to be.

It may be objected to this that one cannot allow a range of actual facts about ordinary objects to be grounded in modal facts about them, and that anyway, some of the modal facts in question, laws of nature, are formulated for entities—persistents—of the very kind for which the difficulties with which we are grappling arise. But both these points can be met by pointing out that the objects for which the difficulties arise are rather special, in that they are homogeneous: this is required to ensure that the region of space occupied by a given part at the beginning of an interval in a case of motion is *continuously* occupied by an object of *exactly* the same dimensions as that which was there initially. And we might hold that for this special category of object (only) transtemporal identity is grounded in modal facts in addition to the continuity considerations embodied in (C). The view that homogeneity gives rise to a *conceptually* special case is at least worthy of consideration.

There is also another way in which Kripke's example might be conceptually special. Perhaps the relevance of the dispositional facts is limited to the question of how it can be told whether or not a homogeneous sphere is rotating, yet the facts about the identity conditions of the half-spheres can still be grounded: grounded in facts about the identity of *other* objects. The obvious response is that the same problems will arise for the other objects, but this response assumes that we never reach a level on which it does not make sense to say that the objects in question have parts; for the problem only arises when we consider parts of things. However, 'simple' objects, or 'atoms', are by definition objects which do not have parts; this means, e.g., that the description 'the portion of atom a in region r' does not denote, if region r is a subregion of the region occupied by a. For atoms, at any rate, criterion (C) is a complete account of transtemporal identity; but then we can ground the transtemporal identity of the half-spheres in that of the atoms which make them up. To this it will be protested that the matter of the sphere need not be composed of atoms (is not, if this is implied by homogeneity): but this is just to say that the case is conceptually special, to the extent that our ordinary concept of matter is that of something composed of atoms. *A priori* atomism has rather sunk from view with the percolation of science into common knowledge, but the infinite divisibility of matter is not a hypothesis with which thinkers have been instantly comfortable, merely waiting for science to confirm or disconfirm it. This, then, would be another area of scope for manoeuvre with Kripke's case.

In conclusion, it should be pointed out that there is no pressing need for us to pursue attempts to undermine the *prima facie* Haecceitist moral of the example very much further, for the theses we are advancing depend only on the correctness of the doctrine that facts about transworld identity and non-identity must be intrinsically grounded. We have seen how the transtemporal facts in the case of the sphere manifest themselves dispositionally, and how a difference in causal antecedents may also be relevant to our grasp of the difference between the case of rotation and the case of non-rotation. This puts the example in a different class from the earlier examples of bare transtemporal identities, so that a counterexample to the sufficient conditions of transworld identity for sets and organisms we have endorsed, to be convincing, would have to have features like those of Kripke's case. But there is a difficulty in principle here: causal influences do not cross possible worlds, and dispositional facts are already modal facts. That is, there is nothing available in terms of which the identity in an alleged counterexample to our transworld sufficient conditions could manifest itself, or could come about. Such an identity would have to have the mysterious ineffability characteristic of alleged examples of genuinely bare identities, and we have seen no reason to take any such case seriously. Thus the doctrine on which we have based our defence of the necessity of origin stands.

7

Fuzzy Essences and Degrees of Possibility

1. *Two paradoxes*

IF WE were to confine our attention solely to the cases of sets and their members, and organisms and their propagules, we would be encouraged to generalize from Membership Rigidity and the Necessity of Origin that it is essential to any non-simple object to come from those entities which it in fact comes from, or which it comes from in an arbitrary world. However, this would be incorrect even for the case of certain organisms, ones which come from cells which do not function like propagules.[1] The *slime mould* is a tiny slug which is formed from the fusion of many single and largely indistinguishable amoebae.[2] Each amoeba exists as a separate, independent individual for a while, reproducing by ordinary mitosis, but when enough are gathered together in one place, they assemble themselves together into a single organism which is not just a mere collection of amoebae, but rather a functionally differentiated creature which leads a life of its own. The trouble with the proposed generalizations of (MR) and (K) is that they would imply that each individual amoeba is essential to whichever slime mould it becomes a part of, but there is simply no intuition that any such relationship obtains. In advance of philosophical argument, most people would be uninclined to deny that a given slime mould could have been formed from a slightly different collection of amoebae; on the other hand, there would be much less agreement that a given slime mould could have been formed from a completely, or even very, different collection; and those who have these intuitions must therefore say that no one constituent amoeba of a slime mould is essential to it, and yet some kind of essentiality of origin attaches to a sufficiently large proportion of these amoebae.

[1] A propagule directs the development of an organism.
[2] See Ede [1978, pp. 9–14].

Artefacts yield a more familiar example of the same phenomenon. A fairly complex artefact, such as a watch, is made from a variety of components according to a particular design, and it is not very plausible to insist that each of a given watch's parts is essential to it, or that it could not have had a slightly different design. On the other hand, it does seem plausible to say that it could not have differed considerably in design or in the parts which make it up: then it would not be *this* watch any more. So what we are encountering here is a certain *vagueness* in the individual essences of entities which are made up of parts and constructed according to particular specifications. To specify the essences of such entities, we need to find some way of representing the thought that if an entity of this sort is made up (without leftovers) of parts from a given set, then as we consider sets of parts which have less and less in common with the given set, it becomes *less and less possible* for the entity to have been constructed from the set under consideration. In effect, then, we must find a way of introducing *degrees of possibility*.

There is more at stake here than merely a question about the scope of essentialist principles like (MR) and (K). In our remarks above, we have endorsed what we might call a *tolerance* principle about the *haecceity* or *thisness* of an artefact (of course, our use of the term 'haecceity' does not indicate agreement with Haecceitism). A general formulation of tolerance with respect to the parts of which an artefact is made is this:

(T) Necessarily, any artefact could have originated from a slightly different collection of parts from any one collection from which it could have originated.

The intuitive justification for the form of (T) is as follows. First, although we agree (let us assume) that, in fact, the same artefact could have been made from slightly different parts, we do not believe that there is some special property of actual artefacts or the actual world which makes this so: even if things had been different, and artefacts different from the actual ones had existed, there would still have been this tolerance. Hence the initial 'necessarily' in (T). Secondly, the formalized version of (T) will contain a conditional with antecedent and consequent each governed by '$\Diamond$', since the effect of (T) is to say that *if* some make-up is a possibility for some artefact, *then* some very slightly different make-up is also a possibility: if α is a possible artefact, then the schematic form of an instantiation of (T) is

$$\Box[\Diamond F\alpha \rightarrow \Diamond G\alpha].$$

This expresses the idea that the ground of truth of (T) lies wholly in the smallness of the quantity of change being contemplated; of course, this is only strictly true under the simplifying assumption that the 'importance' of a given part to an artefact is not weighted.

But however natural (T) appears to be, it is easy to see that it is in some tension with our doctrine that facts about identity must be intrinsically grounded, the doctrine upon which our defence of the essentialist theses of the previous chapters was based. For it is possible to use (T) to provide an apparent proof that there can be both bare transworld identities and bare non-identities.

The argument for bare identity is due to Chisholm, so we call it Chisholm's Paradox (it is this argument to which Quine is referring in our quotation in §2 of Chapter 3, where he contrasts transworld and transtemporal identity on the ground that anything can be changed to anything by easy stages through some connecting series of possible worlds).[3] Let $\langle w_1 \ldots w_n \rangle$ be a sequence of worlds and let $\langle \alpha_1 \ldots \alpha_n \rangle$ be a sequence of artefacts such that each α_i exists in w_i, each α_i is constructed according to the same specifications and no α_i changes its parts through time (for the sake of simplicity, these last two conditions will be in force until further notice). Next, suppose that but for a very few components, each α_i is made from the same parts as α_{i+1}, yet the members of the pairs (α_i, α_{i+1}) differ from each other in such a way that as i increases so the number of parts α_i has in common with α_1 decreases, until we reach α_n, which has no parts in common with α_1. This set-up is a model of certain possibilities allowed by the tolerance principle: w_2 may be taken to be a world which realizes the possibility that α_1 is made of such and such parts, those which make up α_2; that is, $\alpha_1 = \alpha_2$. But then w_3, which by (T) may be taken to realize the corresponding possibility for α_2, thereby realizes a possibility for α_1, and so on, until we reach the conclusion that w_n realizes a possibility for α_1. But α_n is made of completely different parts from α_1, so this gives us our example of an identity which is a bare identity.[4]

We can also give an exposition of the paradox without appeal to possible worlds. For each w_i, let ϕ_i be a predicate which says with rigid designators what parts α_i is made of in w_i, and let us replace 'α_1' with 'α' and treat w_1 as the actual world. Then the following is a classically

[3] See Chisholm [1968].

[4] Bare, that is, relative to the simplifying assumptions of the case. A stronger case is obtained by allowing small changes in design from world to world as well.

sound argument, for its conditional premisses are true by the tolerance principle, the minor premiss is true since '$\phi_1(\alpha)$' is actually true, and the only rule of inference employed is *modus ponens* (→-Elimination):

$$\begin{array}{l} \Diamond\phi_1(\alpha) \\ \Diamond\phi_1(\alpha) \rightarrow \Diamond\phi_2(\alpha) \\ \quad\vdots \qquad\quad \vdots \\ \underline{\Diamond\phi_{n-1}(\alpha) \rightarrow \Diamond\phi_n(\alpha)} \\ \Diamond\phi_n(\alpha) \end{array}$$

So α could have been constructed from parts none of which feature in its actual construction. This gives us bare identity and thus a contradiction with our defence of (MR) and (K). Note also that, in resolving this paradox, we will be defending *de re* modality against Quine's criticism that you can change anything to anything by easy stages through some connecting series of worlds, for we will show that our doctrines about identity are consistent with the phenomenon of tolerance in the thisness or haecceities of certain sorts of things.

The argument for bare facts about transworld differences employs the same resources as Chisholm's Paradox.[5] Our intuition of tolerance in the haecceities of artefacts went along with an intuition that one and the same artefact could not be made from entirely different sets of parts in different worlds. According to this latter intuition, α_1 and α_n are indeed distinct things, so let this be granted. However, by the tolerance principle, there is a sequence of worlds σ_1 like the first half of $\langle w_1 \ldots w_n \rangle$ and a sequence σ_2 like the second half, only in the reverse order, beginning with w_n, each sequence terminating in a world just like a particular world, say w_k, from the middle of the original sequence, such that in the last world of σ_2, α_n is just like α_k in w_k, and in the last world of σ_1, α_1 is just like α_k in w_k. Since α_1 and α_n are distinct, so are these worlds, but the only difference between them is in the identity of the artefacts they contain, and the difference between those artefacts is itself an ungrounded difference. So this example has the schematic form of the Eiffel Tower example (Case 1 in §4 of Chapter 5) which we gave as an illustration of an unacceptable drawing of distinctions; hence, if the argument just given is a good one, we were wrong to

[5] This argument is highlighted in Salmon [1979], where it is called 'The Four Worlds Paradox'.

regard that distinction as unacceptable. We call this paradox the Four Worlds Paradox, the four worlds being w_1, w_n, the last world of σ_1, and the last world of σ_2.

For a formulation of the paradox with modal operators, let us introduce the name 'β', whose reference is fixed by the description 'the artefact which would have resulted if . . .', completing the description by filling in the details of α_n's construction in w_n.[6] We have just agreed that β would not be α, but we can also give two arguments like Chisholm's Paradox, one which concludes '$\Diamond\phi_k(\alpha)$' and the other '$\Diamond\phi_k(\beta)$'. Recalling that we are just now holding design constant, the truth of these two statements together with that of '$\Box(\alpha \neq \beta)$', delivered by the necessity of identity, is inconsistent with the requirement that there must be something in virtue of which transworld differences obtain when they do.

2. *Sorites paradoxes*

The two modal paradoxes are Sorites paradoxes, that is, paradoxes of vagueness. This is especially easy to see in the case of Chisholm's Paradox, which is exactly like familiar Sorites paradoxes such as the Paradox of the Tall Man. Corresponding to the tolerance principle (T) for transworld heirlines, we have a tolerance principle for height classification: someone only marginally shorter (taller) than a tall (short) man is himself tall (short). To be absolutely precise, one tolerance principle concerns the application conditions of a single predicate, 'is tall', whereas the other tolerance principle is really a family of principles, one for each artefact, and the role of the various men in the Tall Man is played, for each artefact, by the various actual and possible sets of parts from which an artefact of that design could be constructed. If α is an artefact, then the predicate whose application conditions are tolerant is a predicate expressing the thisness or haecceity of α, for which we may simply use the predicate '$\xi = \alpha$'.

The analogue to the Four Worlds Paradox is obtained by starting with a man five feet in height and applying the tolerance principle to conclude that a man of five foot six is short, and then by taking a six foot tall man and concluding from the tolerance principle that a man of five foot six is tall. Here we have a bare difference in height classification: there is no difference between such men in which their difference

[6] For a general account of descriptive names, see Evans [1979].

in height status consists and, in particular, there is no difference in their height. This kind of bare difference is indisputably ludicrous.

The classification of the modal paradoxes as Sorites paradoxes makes it desirable that the method of resolving them be an instance of a general strategy for resolving Sorites paradoxes. This immediately eliminates some proposed solutions. For instance, it is tempting to think that what is wrong with the argument of Chisholm's Paradox is that, as things actually are, the later worlds in the sequence do not represent possibilities for α_1, but if things had been as they are in w_4, say, then some of those later worlds would have represented possibilities for α_1. Hence it might be suggested that we can resolve the paradox by introducing an accessibility relation, on which some later worlds in the sequence which are accessible to w_4 are not accessible to w_1. But even without examining how the details of this proposal would be worked out, we can see that such a solution is quite *ad hoc*, and does not address the underlying source of the paradox, unless there is a sense in which tall men are not accessible to short men but are to men of medium height. Moreover, the accessibility solution applied to the Four Worlds Paradox entails that the last world of σ_1 is accessible to w_1, but the last world of σ_2 is not. It therefore requires us to distinguish between those worlds, a distinction which has the same problematic status as the one between their contained artefacts. So someone motivated to seek a solution to the Four Worlds Paradox because he does not wish to draw distinctions which mark no differences, could not be content with such a treatment of it.[7]

A short way with the modal paradoxes is simply to deny that the tolerance principle (T) is true.[8] However, this solution again fails the

[7] The same difficulty afflicts an accessibility solution of Chisholm's Paradox. Suppose we say that w_n is not accessible to w_1 and hence that '$\Diamond\phi_n(\alpha)$' is false at w_1, since there is then no world accessible from w_1 in which α satisfies 'ϕ_n'. Nevertheless, it is clearly possible, relative to w_1, that something be ϕ_n, even if not α. More particularly, it is possible, relative to w_1, that things be exactly as they are in w_n but for the identity of whatever satisfies 'ϕ_n', and thus there is a world v, just like w_n but for the mere identity of a single object, a world which, unlike w_n, is accessible from w_1. But when we allow copies of worlds to multiply like this, we manufacture bare facts about the identity relation of the very kind we set out to remove by seeking a solution to the paradoxes. In [1981] Salmon presents a version of the accessibility solution which recognizes our point that Chisholm's Paradox is a paradox of vagueness by classifying the worlds in the Chisholm sequence in three ways, as either determinately possible (determinately accessible) relative to w_1, or as determinately impossible, or as neither. But this is no improvement on a two-valued solution which does not recognize the intermediate cases, since it is still saddled with the distinction between w_n and v.

[8] In Anil Gupta's book [1980] half a chapter is devoted to the modal

test for being non-*ad hoc*, since its analogue for standard paradoxes is quite inadequate. Michael Dummett and Crispin Wright have convincingly argued that the tolerance in the application conditions of such predicates as 'is tall' is a consequence of the point of using them, which is to effect classifications of objects just on the basis of how those objects look. To give up the tolerance principles here would be to eliminate predicates with such a use from our language, since 'sharpening' such predicates would change their use radically. Predicates which are applied just on the basis of how things look cannot have strictly delimited ranges of application, because, as Wright puts it, 'if the conditions under which a predicate applies are to be generally memorable, [that predicate] cannot be unseated by changes too slight to be

paradoxes. Gupta argues (p. 103) that they are not paradoxes of vagueness since genuine examples of the latter can be blocked by sharpening the tolerance of vague predicates in a way which does not block the modal paradoxes. But his reasoning appears fallacious. He points out that if we just stipulate that a man is not bald iff he has 10^6 hairs or more on his head, then Sorites reasoning will not show that every man is bald. This is correct. But next, he claims that no analogous stipulation will block Chisholm's Paradox and, to demonstrate this, chooses the stipulation that necessarily an artefact can have at most one different part at its origination. However, although it is correct that *this* stipulation does not block Chisholm's Paradox, Gupta has simply failed to compare like with like, for this stipulation is not analogous to the one about the number of hairs. On the latter stipulative solution, there are two possible states of the head, bald and non-bald, and the stipulation assigns each possible quantity of hair to one or other state, no quantity being assigned to both; but Gupta's stipulation about artefacts imposes no limit on the possible original constitutions for a fixed artefact α, since we can arrive at any constitution through a sufficiently long sequence of worlds. A stipulation in the modal case with the same effect as the one about baldness will assign each possible set of parts from which it is possible to construct an artefact according to α's design, to one of two sets, the members of the first set being those belonging to the haecceity of α, the members of the second those not belonging to it. In other words, we just stipulatively list the possible original constitutions of α in such a way that no set of parts is both on the list and on its complement (just as no quantity of hair belongs both to the bald state and the non-bald state). Then one of the conditionals in the premisses of Chisholm's Paradox will be false. Thus, if one chooses a stipulation for the modal case which really is analogous to Gupta's one about baldness, a perfect parallel is preserved. Having mistakenly concluded that Chisholm's Paradox is not like a standard Sorites paradox, Gupta then goes on to offer a strange solution to it (pp. 104–7). His idea is that the truth of a transworld identity judgement of the form 'a is the same F as b' depends upon the world with respect to which the judgement is made; so, in Gupta's framework, *meta*language identity judgements are relativized to the indexes of the model, so that which judgements are true will depend on the point of view of the world at which they are evaluated. In so far as I understand this suggestion at all, I read it as a rather inappropriate way of expressing some kind of counterpart-theoretic notion, with worldbound individuals and a non-transitive counterpart relation.

remembered'.[9] In other words, sharp observational predicates would be *unlearnable* if the phenomena to which they apply form a sensible continuum, as do colours, sizes, and so on. Another of Wright's examples involves predicates for stages of human life, such as 'infant', 'adolescent', etc. One who is an infant at time t is still an infant a few seconds later, but then no one ever reaches adulthood. Here the explanation of the tolerance is that with different stage classifications go explanatory distinctions and differences of moral and social status which a sufficiently small degree of development is too slight to support. Hence, in Wright's irresistible illustration, if we are forced to draw a sharp line, as we are in the matter of electoral qualifications, we do so 'with a sense of artificiality and absurdity'. And although it can hardly be used as an uncontested example, the predicate 'person' or 'bearer of a right to life' is surely another case, definitely applying to teenagers and definitely not applying to embryos, and tolerant because small degrees of biological and psychological development cannot constitute the difference between a case in which they do and a case in which they do not apply, while large degrees of development do constitute such differences.[10]

[9] Wright [1975, p. 337]; see also Dummett [1975].

[10] Wertheimer [1971] tries to defend the conservative position about abortion from this suggestion. He writes (p. 81): 'The conservative points . . . to the similarities between each set of successive stages of fetal development . . . if this were the whole conservative argument . . . it would be open to the liberal's *reductio* . . . which says that if you go back as far as the zygote, the sperm and egg must also be persons. But in fact the conservative can stop at the zygote; fertilization does seem to be a non-arbitrary point marking the inception of a particular object.' This is a poor argument; by completely parallel reasoning, we could start with a 10-year-old (by analogy, highly-developed foetus) who is uncontroversially a child (by analogy, person) and appeal to similarities between successive stages of development to conclude that the 70-year-old (by analogy, zygote) is a child, but resist the conclusion ('the liberal's *reductio*') that the corpse is a child, on the grounds that brain death does seem to be a non-arbitrary point marking the ceasing to exist of a particular object. Of course, Wertheimer could reject this analogy on the grounds that 'human being' is not a phase sortal, and so not analogous to 'child'. But what is the relevance of this difference to the claim that the analogy is a bad one? So far as I can see, it would only be relevant if it is supposed that ultimate sortals must have sharp ranges of application. Yet this supposition, if it is not justified on quite independent grounds, just begs the question, since the only candidate for a sharp beginning to the range of application is conception, and that a zygote is a human being is what Wertheimer is trying to *prove*; so he could not appeal to such a doctrine about ultimate sortals without circularity. Wertheimer goes on to say, 'It needs to be stressed here that we are talking about life and death on a colossal scale . . . so the situation contrasts sharply with that in which a society selects a date on which to confer certain legal rights.' But there is no contrast in any respect which tends to show that Wertheimer's reasoning

The distinction between what is possible and what is impossible for an object is as large a distinction as that between the tall and the short, one primary colour and another, or persons and non-persons, and therefore cannot turn upon a small degree of change in the respect relevant to making the difference. For artefacts, which respects are relevant is certainly open to dispute; one account might be that these respects are simply constitution and design, but it is arguable that we should add function to the list, and the period of time during which the artefact exists, and in the rather special case of works of art, perhaps also the identity of the creator. But we shall not try to argue for the correctness of any single account at this point, since the problem which faces us is to develop an apparatus which permits the appropriate respects, *whatever* they may be, to contribute to the individuality of an artefact in such a way that it can survive small amounts of change in these respects. For here is a substantial difference between artefacts and, say, sets, manifested by there being nothing corresponding to intraworld extensionality ((□E) in §2 of Chapter 5) for artefacts. Sameness of parts is not sufficient for identity of artefacts at a world, since the very same parts may turn up at different times as the parts of artefacts with different designs and functions.

Another reason to insist upon the legitimacy of the idea of tolerant haecceities for artefacts appeals to intraworld transtemporal phenomena. One who would allow an artefact to survive replacement of a part within a world must allow transworld tolerance in original constitution, again on pain of laying down a sharp boundary on inappropriate conceptual terrain. Consider a sequence of worlds in which the time at which a particular part of an artefact is replaced by a certain new part is moved further and further back until we have a world in which the artificer is choosing which of the two parts to put in place in his original construction (we hold constant the stretch of time occupied by the lifespan of the artefact): it would be unmotivated to draw the line at this last world, admitting all the others. Someone might try to motivate such a line by insisting that until the artefact's construction

does not fallaciously exploit the vagueness of a concept, as standard Sorites paradoxes do: an insane tyrant could turn the possession of *any* property which is in fact fuzzy into a matter of life and death. It is dismaying, if not surprising, that sophistries such as these influence certain of our legislators. Without recommending any particular stance on the abortion issue, what can be said uncontroversially is that where great importance attaches to how we draw a line which we are forced to draw by certain practical or moral pressures, the best we can do is to see to it that the line is drawn in such a way that no case which clearly should be on one side is on the other, or even close to the edge.

is complete, it does not exist, but then he has to say that when a part is removed, the artefact goes out of existence, since otherwise we would have a bare non-identity across time: just before the last part is put in place, a certain artefact does not exist, while once in place it does, so if that part is then removed again, we have intrinsically indistinguishable entities, and if one is not identical to the complete artefact, neither is the other. Thus someone who admits temporal but not modal tolerance, and tries to motivate his position, has to say that in having a part replaced an artefact goes out of existence and then comes back into existence; according to this legislation, it is strictly false to say of any artefact that *it* is (now) missing a part. Such a view is rather pointless, since the difficulties which arise from modal tolerance have their temporal analogues (see n. 25), and if they can be handled in the temporal case, it seems unreasonable to refuse to extend the solution to the modal one.

We therefore conclude that the tolerance principles underlying the modal paradoxes are as inviolable as those underlying any Sorites paradox, and turn to the problem of extending the most reasonable solution of the standard paradoxes to these modal ones.[11]

3. *The semantics of vagueness*

How should we resolve standard Sorites paradoxes? It is very plausible that such paradoxes arise from the application of a semantic apparatus appropriate only for sharp predicates to languages or portions of natural language containing vague predicates. This view is in conscious opposition to the idea that vagueness arises from deficiency of meaning[12] or is a source of incoherence;[13] rather, vague concepts are held to be legitimate and unproblematic as they stand, so long as we associate only the appropriate semantics with them.

The crucial notion of this semantics is that of the *degree* to which an object falls under a concept, or the degree to which a predicate applies to an object, and there is a familiar tradition of semantics for vagueness

[11] Christopher Peacocke pointed out to me that the line of reasoning here does not yield the same objection to sharpening haecceity predicates as was made to the proposal to sharpen observational or stage predicates, that such sharpening would nullify the point of having the predicates in the language. The conclusion he draws from this is that not every predicate of degree is one to which Wright-like considerations apply.

[12] This is the viewpoint of Fine [1975].

[13] This is the conclusion drawn by Dummett [1975] and, according to Wright also, is a possible moral of the paradoxes.

using this notion, which finds its perhaps most sophisticated and comprehensive formulation in J. A. Goguen's logic for inexact concepts.[14] We have already used the notion of degree above—indeed, it is hard to avoid its use in setting up examples of situations to which we are going to apply vague predicates—and we can formally introduce the notion from the use of vague predicates in the comparative form. Thus, if one of two tall men is taller than the other, then the first is tall to a higher degree than the other, and so satisfies 'tall' to a higher degree than the other. Since satisfaction of predicates transforms into truth of sentences, it follows that the claim that the first man is tall has a higher degree of truth (is more true) than the claim that the second is. It is hard to find a well-motivated objection to any of these transitions, although it must be borne in mind that the resulting notion of the degree to which a person is tall is non-observational (cannot be told just by looking), unlike the question of his general height status.[15]

The suggestion is, then, that the familiar two-valued semantics be modified by including between its two values of absolute truth and absolute falsehood a range of intermediate degrees of truth, each of which is a possible semantic value for a sentence containing a vague expression. Since ordinary cases of vagueness often arise out of sensible continua, it seems reasonable to allow the degrees of truth to form a continuum; to begin with, then, the closed interval [0, 1] on the real line is a useful model of the set of degrees of truth, with 0 playing the role of absolute falsity and 1 the role of absolute truth. A model for a countable propositional language will therefore consist in an assignment of exactly one degree of truth from [0, 1] to each sentence letter, and the truth-value of any sentence can be computed as soon as we generalize the truth-tables for the connectives to the new degree-theoretic framework.

Noting that in two-valued logic a disjunction takes the better of the two values of its disjuncts, and a conjunction the worse of its conjuncts, we obtain the following clauses for degree-theoretic semantics:

(i) $\mathrm{Val}[A \vee B] = \mathrm{Max}\{\mathrm{Val}[A], \mathrm{Val}[B]\}$

and

(ii) $\mathrm{Val}[A \,\&\, B] = \mathrm{Min}\{\mathrm{Val}[A], \mathrm{Val}[B]\}$.

In two-valued logic, the value of a negation is the complement in the two-membered set {1, 0} of the degree to which the negated sentence

[14] See Goguen [1969].

[15] See Peacocke [1981, p. 125].

falls beneath absolute truth. So for negation, we put

(iii) $\text{Val}[-A] = 1 - \text{Val}[A]$.

Clauses (i), (ii), and (iii) together give us the usual interdefinabilities of '&' and 'v' for the truth-value interval $[0, 1]$. But the natural clause for '→' does not preserve its classical definability by the other connectives. Intuitively, we want the conditional to be material in a generalized sense, that is, it should be true if the consequent is at least as true as the antecedent, but we also want it to take values in the other cases which *reflect the gap* in degree of truth between antecedent and consequent. If the antecedent is only marginally more true than the consequent, the conditional should be only marginally less than wholly true, while if the antecedent is much more true than the consequent, the conditional should be considerably less than wholly true, with the limiting case being that of classical falsehood. The simplest clause which bestows these features on '→' is:

(iv) $\text{Val}[A \rightarrow B] = 1 - (\text{Val}[A] - \text{Val}[B])$ if $\text{Val}[A] > \text{Val}[B]$

$= 1$ otherwise.

For this system of propositional logic, we define a formula to be valid iff its value is 1 on any assignment of degrees of truth to its sentential letters, and we say that an argument is valid iff there is no assignment of degrees of truth to its sentence letters such that the value of the conclusion falls below that of the lowest-valued premiss. More precisely, we say:

$$\Sigma \models A \text{ iff } \text{Val}[A] \geqslant \bigwedge \{\text{Val}[\sigma] : \sigma \in \Sigma\}$$

where the expression after '⩾' denotes the greatest lower bound of the values of the members of Σ, relative to the standard order of the reals (there may be no such thing as 'the' lowest valued premiss if there are infinitely many premisses).

Clauses (i)-(iv) suffice for a resolution of standard Sorites paradoxes. Let $\langle a_1 \ldots a_n \rangle$ be a sequence of men of increasing height such that the statement that a_1 is short is wholly true and the statement that a_n is short is wholly false, although there is only a marginal difference in height between adjacent men in the sequence. The tolerance of 'is short' implies, with respect to the two-valued framework, that each conditional of the form:

$$a_i \text{ is short} \rightarrow a_{i+1} \text{ is short}$$

is true. Hence the following argument is classically sound:

$$\begin{array}{l} a_1 \text{ is short} \\ a_1 \text{ is short} \rightarrow a_2 \text{ is short} \\ \vdots \qquad\qquad\qquad \vdots \\ a_{n-1} \text{ is short} \rightarrow a_n \text{ is short} \\ \hline a_n \text{ is short.} \end{array}$$

But this is inconsistent with the fact that a_n is, say, six foot six. However, on the degree-theoretic framework, we see the argument in a different light. The problem is that →-Elimination is an unreliable rule of inference in this framework, in a way that &-Elimination, for instance, is not: if $\mathrm{Val}[A \rightarrow B] = 1$ then applications of →-Elimination are unproblematic, but in our argument none of the conditionals is wholly true. In each, the degree of truth of the antecedent is marginally higher than the degree of truth of the consequent because each a_i is marginally shorter than the corresponding a_{i+1} (note, again, that even the comparative facts about degrees of truth need not be accessible to simple looking and seeing, since a marginal difference in height need not be observationally detectable). By clause (iv), therefore, each conditional is very slightly less than wholly true, and →-Elimination is being used to detach consequents whose degrees of truth are dropping steadily towards 0. The paradoxical argument is therefore a concrete illustration of a possibility implicit in the semantics, that from an absolute truth we may reason through a chain of conditionals each of which is almost wholly true and yet end up with a complete falsehood.

This is an elegant and appealing diagnosis of paradoxes of vagueness; it is because the conditionals are almost wholly true that the argument seems to us to be irresistible, and so, besides being neutralized, its persuasive force is explained. The only serious objections to this approach to vagueness involve what Fine has called 'penumbral connections', which, if they obtain, are inconsistent with the fact that on clauses (i) to (iv), the degree of truth of a compound formula is always the same function of the degrees of truth of its component subformulae: in place of classical truth-functionality, we have degree-functionality. But Fine has given a putative counterexample to degree-functionality, involving a conjunction whose degree of truth is claimed by Fine to be different from what clause (ii) says it should be.[16] 'is pink'

[16] Fine [1975, p. 26].

and 'is red' are contraries (a penumbral connection), and hence, according to Fine, 'α is pink and α is red' must be wholly false. But if α is poised exactly midway between paradigm pink and paradigm red, then each conjunct has a middle degree of truth, approximately 0.5, which, if clause (ii) is to be believed, is passed on to the whole conjunction. Thus, Fine concludes, clause (ii) gives the wrong result in this kind of case.

However, this objection is unconvincing. To say that 'is pink' and 'is red' are contraries, from the degree-theoretic point of view, is not to say that nothing can be both, but rather, to say that nothing can be wholly pink and also wholly red; but a thing can of course be red to a certain degree and pink to a concomitant degree. If in the situation of the example one man says 'α is pink' and another 'α is red' and neither is judged to have uttered something wholly false, why should this fate befall the first man if he anticipates and utters the second man's thought as well as his own, using 'and' to avoid an unnatural break in his speech? A reply of this kind can also be made to someone who holds that 'α is red and α is not red' should be wholly false, or that 'α is red or α is not red' should be wholly true. One reason (not Fine's) for ascribing these truth-values should certainly be rejected: someone might think that further investigation of an intermediate case of red would reveal whether or not the thing in question is *really* red. But this is quite confused; vagueness is an ineliminable feature of colour concepts, and not the product of some discriminatory limitation to which our sensory apparatus is subject.

If the conditional premisses of a Sorites argument are not wholly true, what of the tolerance principles which justify those premisses? Such principles are universal quantifications to the effect that if one thing is related thus and so to another then the second has a certain property if the first has it. So to be precise about the truth-values of these principles we have to extend the degree-theoretic apparatus to quantifiers, i.e. to first-order logic. This will have the additional advantage of enabling us to see how the degree of truth of an atomic sentence such as 'a is short' is determined by the semantic properties of its constituent name and predicate. Our intuition was that a predicate like 'is short' is satisfied by different objects to different degrees, so as to specify its extension, we need to state not just which objects it applies to, but also, for each such object, the degree to which it applies. Following Goguen, we think of such an extension as a function from a set X of objects into the set J of degrees of truth (such functions are sometimes called fuzzy sets, since they can be regarded as determining a set whose members belong to it to different degrees—the set of short things would be an example). Note that, on this approach, vagueness resides entirely in concepts. The objects in X are perfectly determinate and the fuzzy sets themselves also have exact identity conditions: sets of this sort are the same iff the same things are members of each to the same degree.

More generally, if F is an n-place atomic predicate then we assign to F a function χ_F from a set X of n-tuples of objects drawn from a domain D into a set J of degrees of truth (X is called the 'universe' of F). Then for atomic sentences we have:

(v) $\text{Val}[F(t_1 \ldots t_n)] = \chi_F(\langle \text{Ref}(t_1) \ldots \text{Ref}(t_n) \rangle)$.

Existential and universal quantification can always be thought of as equivalent to infinitary disjunction and conjunction, so the quantifier clauses involve the infinitary analogues of Max and Min, the least upper bound (lub) and greatest lower bound (glb) operations:

(vi) $\text{Val}[(\exists v)Av] = \text{lub}\{\text{Val}[A(\underline{a}/v)] : \text{all a in D}\}$

and

(vii) $\text{Val}[(\forall v)Av] = \text{glb}\{\text{Val}[A(\underline{a}/v)] : \text{all a in D}\}$.

Returning to the tolerance principles, we see that none of them is wholly true; for instance, a version of the principle for the Tall Man is

$(\forall x)(\forall y)(\text{Short}(x)\ \&\ \text{y is one centimeter taller than x} \rightarrow \text{Short}(y))$.

Any instance of the conditional matrix with 'a' for 'x' and 'b' for 'y' is either wholly true (because b is not one centimeter taller than a or because 'Short(a)' and 'Short(b)' are both wholly true or wholly false) or else slightly less than wholly true, because 'Short(b)' is slightly less true than 'Short(a)'. Thus, by (vi), the value of the universally quantified sentence is slightly less than wholly true, and can be brought closer and closer to absolute truth by taking smaller and smaller differences in height. Hence our earlier insistence that tolerance principles are true requires qualification when we move out of the two-valued framework: they are merely almost wholly true. But this in itself is sufficient to show that it would be absurd to deny them.

4. *Closgs and counterparts*

Our goal is to extend the best resolution of standard Sorites paradoxes to the modal paradoxes, so the next step is to import the degree-theoretic apparatus into the modal logical framework. There are two obstacles to be overcome at this point, one technical and the other philosophical. We take the technical problem first.

(I) When we compare artefacts across worlds, we assess degrees of similarity in at least two respects, constitution and design; in our presentation of the modal paradoxes, we only allowed constitution to vary, but this was an artificial restriction. Suppose we now consider two artefacts β and γ in a world w and ask to what degree they satisfy '$\xi = \alpha$' at w, where α is some actual artefact. Perhaps β is close in design to α but not in constitution, while the converse is true for γ. So with each of β and γ we can associate a pair of numbers, measuring degree of similarity to α in each of the two respects. Yet there does not seem to be any reason why these two numbers have to be resolvable into a single number giving *overall* degree of similarity to α so that β and γ can be compared by that yardstick. But if only pairs of numbers are available, β and γ may be *incomparable* in respect of overall similarity, and in such a case [0, 1] would be an inadequate model of the set of degrees, since it is *totally* ordered by $\leqslant$.

However, this technical difficulty can be overcome. If there is no fact of the matter about which of x or y possesses the greater or lesser degree of similarity to z because the degrees of similarity which they do possess are incomparable in respect of which is the greater or lesser, then the degrees of truth of two atomic sentences of the form 'a is similar to b' may also be incomparable in respect of which is the greater or lesser, and thus [0, 1] fails to model the degrees of truth, since any two numbers in [0, 1] *are*, of course, comparable in respect of which is the greater or lesser. But it is easy to find mathematical objects of which this condition does not hold. Suppose degree of similarity is given by a pair of numbers each of which is in [0, 1], and each of which measures a single aspect of similarity (imagine only two aspects are relevant in the examples under discussion). Then we might consider $[0, 1] \times [0, 1]$, the set of pairs of reals from [0, 1], as a model of the set of degrees of truth, since we can define 'less than or equal to' as:

$$\langle a, b \rangle \leqslant \langle c, d \rangle \quad \text{iff} \quad a \leqslant c \text{ and } b \leqslant d$$

where the symbol '$\leqslant$' on the right-hand-side stands for the normal order on the reals. This is an example of a *component-wise extension* of a relation or operation: if the relation or operation is defined for single objects, then we define its application to n-tuples of these objects in terms of its application to the objects which occupy the j'th positions of each n-tuple, $1 \leqslant j \leqslant n$. And with this object as our model of the set of degrees of truth, we do obtain incomparable degrees; for example, $\langle 2, 3 \rangle$ and $\langle 3, 2 \rangle$ are incomparable in terms of which is the 'lesser', given our definition which extends this notion to pairs of reals. With such a model of the degrees of truth, there is no need to change any of the connective clauses (i)-(vii) above, since the arithmetical operations which they employ can also be extended component-wise to $[0, 1] \times [0, 1]$. Thus clause (iv) for implication,

$$\text{(iv)}\ \mathrm{Val}[A \to B] = 1 - (\mathrm{Val}[A] - \mathrm{Val}[B]) \text{ if } \mathrm{Val}[A] > \mathrm{Val}[B]$$
$$= 1 \text{ otherwise}$$

should be construed by understanding '$\langle a, b \rangle > \langle c, d \rangle$' to hold just in case $a > c$ and $b > d$, and $\langle a, b \rangle - \langle c, d \rangle$ in this case to be $\langle (a - c), (b - d) \rangle$; 1 is of course $\langle 1, 1 \rangle$, the maximum degree of truth.

Obviously, the restriction to two respects of similarity in the above discussion is inessential: for any n, we can admit n respects of similarity and choose $[0, 1] \times \ldots \times [0, 1]$ (n times) as our model of the degrees of truth, extending arithmetical notions component-wise. In particular, we say

$$\langle a_1 \ldots a_n \rangle \leqslant \langle b_1 \ldots b_n \rangle \text{ iff } a_i \leqslant b_i,\ 1 \leqslant i \leqslant n.$$

It turns out, however, that even such sets as finite products of the unit interval have structural properties which are unnecessary to model degrees of truth and to permit the definition of reasonable clauses for the connectives. In algebraic investigations of this question, Goguen has shown that, for a logic of vagueness, the minimum acceptable structural requirement on the set of degrees is that it have the order type of a complete lattice-ordered semi-group ('closg') in which the lattice maximum is identity for the group operation *.[17] Fortunately, to understand what is to come, the reader will be pleased to learn that he need only keep in mind two examples of closgs, $[0, 1]$ as above and $[0, 1] \times [0, 1]$, or perhaps more generally, $[0, 1] \times \ldots \times [0, 1]$. It is therefore unnecessary to pursue formal questions here any further, or to explain the terminology involved in Goguen's acronym.

(II) So much for the technical difficulty in extending degree-theory

[17] Goguen [1969, p. 354].

to modality. The philosophical difficulty concerns the coherence of the notion of the degree to which an object satisfies such a predicate as '$\xi = \alpha$' at a world. In the standard semantics for S5, transworld heirlines for objects are given by real crossworld identities in the model, and thus the only object which satisfies '$\xi = \alpha$' at a world is α. So if there can be degrees of satisfaction of '$\xi = \alpha$' at a world w, then it looks as if there must be degrees of being identical to α at worlds. Yet the notion of degrees of identity is incoherent. We saw that the idea that a predicate F is a predicate of degree is justified by the admissibility of the comparative form 'x is more F than y', but it would be quite hopeless to try to make literal sense of 'x is more identical to α than is y', and we will not waste space in the attempt.

Instead, we need to replace standard S5 semantics with some other sort, in which transworld heirlines are given not by real crossworld identities, but rather by some other transworld relation which it does make sense to regard as a relation of degree. The prescient reader will have anticipated that the Counterpart Theory which we outlined and developed in §4 of Chapter 3 is about to reappear, for as the counterpart relation was originally explained by Lewis, it is a relation whose crossworld extension is fixed by considerations of crossworld similarity. Since there is no problem at all about degrees of similarity, degrees of counterparthood are equally straightforward. And using Counterpart Theory, it will transpire that what is satisfied to different degrees at different worlds is *any de re* modal predication of α, the element of degree being introduced by the modal operator, which in turn introduces the counterpart predicate; in other words, the tolerant concept is not that of being identical to α, but the more general concept of being a possibility or necessity for α, and we shall see that this holds the threat of incoherence at bay.

Nevertheless, although we have already dismissed a number of objections to counterpart theory, our discussion in Chapter 4 left two related issues unresolved. One was the problem of defining similarity so that certain intuitively true modal judgements are ascribed the right truth-value, and the other was the problem of dealing with the necessity of identity. The difficulty we mentioned then is that if we define similarity so as to obtain pleasing consequences, we introduce an element of the ungrounded if the definition is stipulative at any point, and so we fail to meet Quine's challenge to *elucidate* a relation which can be used to make sense of *de re* modality; or else we end up with a completely elucidated relation which is structurally isomorphic to a transworld identity relation, and the difference between counterpart-theoretic and standard semantics collapses.

We are now in a position to see how this dilemma is to be avoided. In connection with artefacts, degree of counterparthood is fixed by degree of similarity in particular respects, which do not include the subsequent history of an artefact, involving, say, who owns it, or the path it traces through space and time. It is thus possible for an artefact to have had a very different subsequent history from the one it actually has, while some other artefact has a career rather like that of the given artefact. So our counterpart relation, which admits of degrees, is completely grounded in facts about crossworld similarity, but only facts of a particular sort. Furthermore, it is in no way isomorphic to identity. This is not merely because it admits of degrees, but because it may be many-one or one-many between a pair of worlds. This fact raises our worry about the necessity of identity (formulae (21)–(23) in Chapter 3, (21) reproduced as (1) below) which will certainly fail in the present system. Consider

(1) $a = b \rightarrow \Box(a = b)$

which translates as

(2) $a = b \rightarrow (\forall w)(\forall x)(\forall y)(Cxaw \,\&\, Cybw \rightarrow x = y)$.

Suppose a and b are the same, and that w is a world where there are *two* artefacts c and d such that c and d are of the same design as a and each has half of a's parts. Other things equal, this makes each a counterpart of a (hence of b) to degree 0.5, and thus the conditional

(3) $(Ccaw \,\&\, Cdbw) \rightarrow c = d$

has degree of truth 0.5, since its antecedent has degree of truth 0.5 and its consequent is wholly false. Hence (1) is invalid on our approach; that is, it is not wholly true on every assignment of degrees.

If (1) is invalid, what becomes of the argument for it from '$(\forall x)\Box(x = x)$' and Leibniz's Law (see §5 of Chapter 3)? In fact, it is not the Law which causes the problems, but the premiss '$(\forall x)\Box(x = x)$'. Consider the sentence '$\Box(a = a)$'. In our evaluation clauses for '$\Box$' and '$\Diamond$' in §4 of Chapter 3, we use 't_i' for the i'th individual constant *token* in an object language expression, and there are two such constant-tokens in '$\Box(a = a)$', even though they are both of the same type. Thus, by evaluation clause (xiv), the truth of '$\Box(a = a)$' at a world requires that every world w, if c is a counterpart of a at w and d is a counterpart of a at w, then 'c = d' is true at w. We have still to reformulate those clauses to allow for degrees intermediate between 0 and 1, but we can already see, in virtue of the example just given, that 'c = d' may be wholly false although it is not wholly false that c is a counterpart of a at w, nor that b is a counterpart of a at w. The same point is perhaps more obvious if we make it in terms of translation rather than evaluation, since the translation of '$\Box(a = a)$' is '$(\forall w)(\forall x)(\forall y)(Cxaw \; \& \; Cyaw \rightarrow x = y)$', which, in the light of our example above, is obviously not a theorem of Counterpart Theory.[18]

Thus while every *de dicto* modal thesis about identity has the same truth-value in the present framework as it has in the classical framework, a difference emerges over the *de re*, *not* because identity somehow becomes fuzzy, but because *de re* sentences introduce a new fuzzy relation, that of counterparthood, which in turn gives rise to degrees of possibility. What then of Kripke's claims that the necessity of identity is intuitively valid? The examples used by Kripke to invoke the intuition that such formulae as (1) are valid usually involve objects, such as planets and people, to which the notion of part has no very natural application. So we can respond to Kripke that when one considers entities of other categories, there may no longer be an intuition in favour of the Necessity of Identity. For instance, let α and β be clocks of identical design on opposite walls of a room and imagine a possible world in which there is only one clock in the room, made out of half of α's parts (strictly, counterparts of these) and half of β's, but with the same design as the actual clocks. Such a state of affairs is evidently genuinely possible and, in virtue of this, the present approach dictates that the judgement 'These two actual clocks could have been a single clock' is not wholly false. However, it is simply untrue that there is a firm pretheoretic intuition that this result is unacceptable.

There is one further objection to the use of counterpart theory which we ought to consider. In Chapter 3, following Hazen, we accused

[18] If we require for the truth of '$\Box(a = a)$' only that at every world w every counterpart of a at w is identical to itself, then what Kripke claims to be an instance of Leibniz's Law,

(i) $a = b \rightarrow [\Box(a = a) \leftrightarrow \Box(a = b)]$

would in fact not be, since by the same principle its translation would be

(ii) $a = b \rightarrow [(\forall w)(\forall x)(Cxaw \rightarrow x = x) \leftrightarrow (\forall w)(\forall x)(\forall y)(Cxaw \; \& \; Cybw \rightarrow x = y]$.

But whether or not a formula is an instance of Leibniz's Law is merely a matter of its form, and (i) most certainly is. Hence the counterpart-theoretic truth-condition (ii) is incorrect.

Kripke and Plantinga of confusing object and metalanguage in objecting that Counterpart Theory misrepresents the contents of modal judgements. Nathan Salmon has attempted to give a more sophisticated version of this objection which escapes Hazen's refutation of it.[19] Salmon's idea is to put the counterpart relation into the modal object language and then to compare the counterpart-theoretic truth-conditions of 'it is possible that a is F' with 'it is possible that some counterpart of a is F'. But there is really no improvement of the Kripke/Plantinga objection to be extracted from this line of thought. Salmon claims that 'intuitively', the second modal sentence,

(4) $\Diamond(\exists x)[(Cxa) \& Fx]$

is weaker than the first,

(5) $\Diamond Fa$

yet, according to Counterpart Theory, (4) entails (5); therefore Counterpart Theory misrepresents the content of (5).[20] But (5) is a logical consequence of (4) only when the two-place predicate 'C' is treated as a logical constant of a certain unfamiliar and highly technical sort.[21] So Hazen's reply to Kripke and Plantinga, that pretheoretic intuitions are not in question, applies here too. Hence all philosophical obstacles in the way of using counterpart theory to resolve the modal paradoxes may be overcome.

5. *Counterpart theory with degrees of possibility*

We now give a brief but rigorous formal description of counterpart theory with degrees, by defining a degree-model M for the language L_c

[19] Salmon [1981, p. 235]. The discussion which follows has benefited from helpful remarks Salmon made in response to some earlier, ineffective, criticisms of his position which I advanced.

[20] The translation of (4) is

$$(\exists w)(\exists z)(Czaw \& (\exists x)(E(x, w) \& Cxaw \& Fxw))$$

and the translation of (5) is

$$(\exists w)(\exists x)(Cxaw \& Fxw).$$

(4) entails (5) because of the condition in Counterpart Theory that each thing is its own unique counterpart at the world where it itself exists and at worlds where it has no existent counterpart.

[21] To legitimize the translation of the two-place predicate 'C' in (4) by the metatheoretical three-place relation of counterparthood, the two-place relation has to have a constant interpretation in every model, in the sense that 'Cab' should hold at a world w iff a is a counterpart of b at w in the model in question.

of counterpart theory, which, in combination with the translation scheme mapping L_m formulae into L_c formulae given in §3 of the Appendix, will enable us to exhibit precisely the invalidity of the modal arguments which constitute the paradoxes. An equivalent way of proceeding would be to give another model theory for L_m, one in which formulae take various degrees of truth other than 1 and 0, and such a theory can be read off from the definition of degree model for L_c; but translations are slightly easier to work with.

The only vague predicate of L_c will be the counterpart predicate, so its extension will be given as a fuzzy set, that is, as a function into the set J of degrees of truth, which will in turn be an arbitrary closg. However, in order to treat all predicates uniformly we will also specify the extensions of the exact predicates as functions. An n-tuple of objects either definitely is or definitely is not in the extension of an exact n-place predicate, and thus the extension of such a predicate can be given by a function which maps its members to the value 1, the maximum of J, and its non-members to the value 0, the minimum of J.

The language L_c contains two sorts of terms, including the constant 'w*' of sort 1, all the constants of L_m, which are of sort 2 in L_c, and for each n-place predicate of L_m an n+1-place predicate whose last place is reserved for a term of a sort 1. So the existence predicate '$E(\xi)$' of L_m is correlated with '$E(\xi, \zeta)$', a predicate of sort $\langle 2, 1\rangle$. In addition, L_c contains a three-place predicate '$C(\xi_1, \xi_2, \zeta)$' of category $\langle 2, 2, 1\rangle$, which is read as 'ξ_1 is a counterpart of ξ_2 at ζ'. The two-sorted language is for ease of readability while, as before, the three-place counterpart predicate is needed to obtain a correct logic of existence.

A degree-model M for L is a two-sorted 9-tuple

$$\langle W, D, J, Q, R, I, H, w^*, v\rangle$$

where W is a set of entities of the first sort (worlds) and D is a set of individuals. J is a closg (see §4) whose elements are K-tuples of real numbers from the interval [0, 1], for whatever fixed k is the number of distinct criteria by which counterparthood is assessed. The lattice ordering is defined componentwise and the group operation * by

$$\text{(i) } \langle a_1 \ldots a_k\rangle * \langle b_1 \ldots b_k\rangle = \langle (a_1 \times b_1) \ldots (a_k \times b_k)\rangle.$$

Q is a distinguished function from $D \times W$ into the subset $\{\underline{0}, \underline{1}\}$ of J, where $\underline{0}$ is the k-tuple $\langle 0 \ldots 0\rangle$ and $\underline{1}$ the k-tuple $\langle 1 \ldots 1\rangle$, absolute falsity and absolute truth respectively. Q interprets the two-place existence predicate of L_c and is subject to the constraint that

(ii) for all w, w′ in W, if $w \neq w'$ then if $Q(x, w) = \underline{1}$ then $Q(x, w') = \underline{0}$.

So we are going to restrict ourselves to worldbound individuals, things which exist in at most one world. However, this restriction is inessential and we will later mention a reason for relaxing it in a more complex semantics.

R is a function from $D \times D \times W$, the universe of the counterpart relation, into J, and meets a number of conditions. First, reflexivity in variables of the second sort:

(iii) for all x in D and w in W, if $Q(x, w) = \underline{1}$ then $R(x, x, w) = \underline{1}$.

But symmetry in individual variables with respect to degree is plausible only when design is held constant; if complex artefacts can be counterparts of simple ones, proportion of parts in common may not be the same. And, obviously, no version of transitivity with respect to degrees is desirable. But we do impose two other conditions:

(iv) for all x in D and w in W, if all y in D are such that if $Q(y, w) = \underline{1}$ then $R(y, x, w) = \underline{0}$, then for all y in D, $R(y, x, w) = \underline{1}$ iff $y = x$.

This says that any object with no existing counterpart at w is its own sole counterpart there, and, it will be recalled, is the condition motivated by the analogy with standard possible worlds semantics, in which an atomic predicate may be satisfied at a world by an object which does not exist at that world; thus on counterpart-theoretic semantics as well as the standard one, the Falsehood Principle is imposed on or withheld from atomic predicates on a case by case basis. We will also insist that counterparthood be properly a crossworld relation when it holds between distinct things:

(v) for all x, y in D and w in W, if both $Q(x, w)$ and $Q(y, w) = \underline{1}$, then $R(x, y, w) > \underline{0}$ iff $x = y$.

I interprets the identity symbol of L_c, which is of category $\langle 2, 2\rangle$ and is a function from $D \times D$ into $\{0, 1\}$ such that $I(a, b) = 1$ iff $a = b$. H is a set of characteristic functions f_i^{n+2}, one for each n+1-place non-logical predicate F_i of L_c. w* is a designated member of W, and, lastly, v is a function which assigns members of D to constants of sort 2 under the constraint that

(vi) $v(c) = x$ only if $Q(x, w^*) = 1$.

However, it is also possible to have names for non-actuals in L_c.[22]

The connective clauses of §3 need to be modified slightly for more general truth-value sets, which, for instance, need not be complemented. To interpret negation and implication, Goguen defines the functions Neg and Imp thus:

(vii) $\text{Neg}(\langle a_1 \ldots a_k \rangle) = \langle (1 - a_1) \ldots (1 - a_k) \rangle$

and

(viii) $\text{Imp}(a, b) = \text{lub}\{x : (x^*a) \leqslant b\}$.[23]

For disjunction and conjunction we take the operations of join and meet defined componentwise in J by Max and Min respectively, while

[22] It may be thought that there is a difficulty in counterpart theory with names for non-actuals such as 'β' from §1, on the grounds that for a modal sentence containing this name to have a determinate truth condition, 'β' must refer to a particular one of the many worldbound individuals with constitution ϕ_n, and it is unclear how it is to be made to do so. But the condition for determinacy of reference is just that it be possible that there is an object such that necessarily 'β' refers to it alone, i.e.

$$\Diamond(\exists x)\Box(\forall y)(R(\text{'}\beta\text{'}, y) \leftrightarrow x = y).$$

In counterpart theory, this is the condition:

$$(\exists w)(\exists x)(E(x, w) \; \& \; (\forall u)(\forall z)(\forall y)(Czxu \; \& \; E(y, u) \rightarrow [R(\text{'}\beta\text{'}, y, u) \leftrightarrow y = z])).$$

When counterparthood is identity-like because criteria for it are as strict as those for transworld identity for sets (two-valued counterpart theory) this condition is clearly satisfied, and thus the reference of 'β' is determinate. This is already enough to show that the worldbound nature of the individuals in counterpart theory is not an obstacle to the introduction of names for non-actuals. When we are in the degree-theoretic framework, the condition remains true only by allowing R to be a relation of degree such that if y and z are distinct and both counterparts of x at w′, then the degree to which 'β' refers to y at w′ is 1 minus the degree to which z is a counterpart of x at w′; this is a strange but not impossible consequence, granted that we are talking about reference to non-actuals. Only if some world where there is an artefact of the fixed design with constitution ϕ were actual would 'β' be a name of an actual. Note that by our views about the conterpart relation, any sentence of the form $\Diamond A(\beta)$ will have a fixed degree of truth regardless of which of the ϕ_n worldbound individuals 'β' is taken to denote, so for evaluation purposes we can just pick one at random. Note also that the modal determinacy condition implies that possibly there is an x such that actually, 'β' refers to x; since this is true, the Falsehood Convention could not be applied to 'refers'.

[23] Goguen [1975, p. 356]. With respect to note 24 below, it is important to note that if subtraction always makes sense, then a component-wise generalization of clause (iv) from §3 also satisfies Goguen's conditions on an implication function.

the quantifier clauses stay the same as in §3, as do the definitions of validity for formulae and arguments. This completes the account of degree-theoretic model theory for counterpart theory.

6. *Consequences*

To begin, let us see how a semantics of degrees in counterpart theory defuses the modal paradoxes. Chisholm's Paradox is straightforwardly dealt with, for when we consider its modal operator formulation in §1, we see that its conditional premisses have L_c-translations of the form:

$$\text{(A)}\ (\exists u)(\exists x)[Cx\alpha u \ \&\ \phi_i(x, u)] \rightarrow (\exists v)(\exists y)[Cy\alpha v \ \&\ \phi_{i+1}(y, v)]$$

and that no instance of (A) is wholly true. Rather, in each such instance, the consequent is slightly less true than the antecedent because anything with constitution ϕ_{i+1} is slightly less similar to α than anything with constitution ϕ_i, and so is a counterpart of α at a world, if at all, to a slightly lesser degree than something with constitution ϕ_i. Our clauses for the connectives ensure that this small gap in degree of counterparthood translates itself up through the structure of the formula to yield an expression slightly less than wholly true: suppose the degree of truth of the antecedent of some instance of (A) is k, in virtue of the highest degree of counterparthood to α open to an object with the relevant constitution at any world u being k (that object has that constitution at u to degree 1, so the conjunction in the antecedent has degree Min(k, 1) = k); then the highest degree of counterparthood to α open at any world to any object with a constitution even further removed from that of α at the actual world will be correspondingly lower, yielding a lower degree of truth for the consequent of the instance of (A); so by clause (iv) for '→', the whole instance gets a degree of truth less than 1. Thus our resolution of Chisholm's Paradox is absolutely parallel to the resolution of the Tall Man.[24]

[24] Some find it natural to formulate the conditional premisses of Chisholm's Paradox as counterfactuals of the form:

$$\phi_i(\alpha) \mathrel{\Box\!\!\rightarrow} \Diamond\phi_{i+1}(\alpha)$$

and although this makes the analogy with the Tall Man more remote, essentially the same points hold. First, to handle translation of counterfactuals into L_c, we need to define a new operation 'Rel*' thus. If there are term-occurrences $t_1 \ldots t_n$ in A outside the scope of modal operators in A, then

$$\text{Rel*}(A, w) = (\exists u)(\exists v_1) \ldots (\exists v_n)(Cv_1 t_1 u \ \&\ \ldots \\ \&\ Cv_n t_n u \ \&\ \text{Rel}(A(v_i/t_i), w));$$

The Four Worlds Paradox is dealt with similarly, since 'β' is a descriptive name which may be assigned a definite non-actual individual (see note 22). We say that the last world of σ_1 and the last world of σ_2, alleged to differ only with respect to the identity of a particular artefact, are in fact the same world, and those artefacts the same artefact. This artefact is a counterpart at its world of both α_1 and α_n, a counterpart of each to the same degree, even though α_1 and α_n are not counterparts of each other at all. In modal language, we can say that the 'Four' Worlds Paradox involves a world which realizes a state of affairs which is semi-possible, the state of α and β being the same. In deriving the paradox, we appealed to the 'fact' that α and β are necessarily distinct, but here we were relying on a version of the necessity of identity,

$$(6)\quad (\forall x)\Box(\forall y)(x = y \rightarrow \Box(x = y))$$

which is, of course, invalid in the present system. Its L_c translation is:

$$(7)\quad (\forall x)(E(x, w^*) \rightarrow (\forall y)(\forall w)(\forall z)(Cyxw \;\&\; E(z, w) \;\&\; y = z \rightarrow (\forall u)(\forall s)(\forall t)(Csyu \;\&\; Ctzu \rightarrow s = t)))$$

and the validity of (7) would require that the counterpart relation be transitive in its individual variables in this sense, that if y is a counterpart of x at u to degree m and z a counterpart of y at v to degree n then z is a counterpart of x at v to a degree not less than Min(m, n). But the relevance of relations of degree to paradoxes of vagueness is precisely

otherwise, Rel*(A, w) = Rel(A, w). We now expand L_c by adding the three-place relation symbol S(u, w, v) for comparative similarity ('u is more similar to w than is v'), a symbol of sort ⟨1, 1, 1⟩, and a degree model is concomitantly understood to be a 10-tuple whose new component is a function S: WxWxW → {0, 1}. The Lewis–Stalnaker analysis of '□→' then motivates the following:

$$\text{Rel}(A \;\Box\!\!\rightarrow B) = (\exists u)(\text{Rel}^*(A \;\&\; B, u) \;\&\; (\forall v)(\text{Rel}^*(A \;\&\; -B, v) \rightarrow S(u, w, v))).$$

Curiously, this renders every counterfactual in the quasi-Sorites argument wholly false, excepting only the very first. The problem lies with Goguen's clause for Imp, which makes every conditional with a wholly false consequent itself wholly false, provided just that the antecedent has some degree of truth, regardless of how little. So his clause, in this case, does not reflect the gap. However, for the purposes of handling our argument, we can use instead a generalized version of (iv) from § 3. Then this counterfactual version of Chisholm's Paradox will not lead us to the paradoxical conclusion, since we will resist the inference from 'Possibly, P' and 'If it had been that P, it would have been possible that Q' to 'Possibly Q' for the familiar reason, that the degree of possibility of Q is dropping off.

that they fail to be transitive. Thus, by rejecting (6), we block the paradoxical argument. But this move is not *ad hoc*. Rather, it is a consequence of our view that the modal paradoxes are Sorites paradoxes, and of our applying a general technique for dealing with such paradoxes to the modal case.

We began this discussion with two aims in view; one was to investigate what kind of essentialism is possible with respect to entities like the slime mould and artefacts, and the other was to demonstrate that the paradoxes pose no threat to our doctrine that facts about identity must be intrinsically grounded. In connection with the first aim, we are now in a position to see that artefacts may be ascribed fuzzy essences. The fuzzy essence of a given artefact may be regarded as a set of pairs, each pair consisting of an object and a degree, the second member of the pair specifying the degree to which the first member is a counterpart of the given artefact. So a fuzzy essence is a perfectly determinate object, like a fuzzy set. In modal operator discourse, two theses are justified by ascription of such fuzzy essences, analogous to the pairs ⟨(MR), (CE)⟩ and ⟨(K), (PSI)⟩. First, at least if each of an artefact's parts is equally important to its functioning, we can say that it is essential to an artefact to have *most* of the parts it actually has or could have had. The logical form of this thesis, which we label (Z), is awkward to represent because of more general difficulties with vague quantifiers like 'Most', but if we write '$Mx\phi x: \Psi x$' to mean 'Most of the things which are ϕ are Ψ', then one possible regimentation is:

$$\text{(Z)} \quad \Box(\forall x)\Box_1(E(x) \rightarrow \Box(E(x) \rightarrow [MyP(y, x): A_1(P(y, x))]))$$

where '$P(y, x)$' is to be read as 'y is a part of x'. Under fairly reasonable assumptions about the interpretation of 'Most', we can even say that (Z) is wholly true. Suppose that b is a counterpart of x to degree k at the fixed world w_1, and that c is a counterpart of x at some world v to degree j. Is it really possible to choose v so that the degree of truth of 'most of the parts of c at v are parts of b at w_1' is so low that subtracted from j, it yields something less than k? In comparing the parts of c at v and the parts of b at w_1, we are again comparing objects some of which are counterparts, and we should expect that, on the intended interpretation, these degrees of counterparthood will be reflected by the degree of counterparthood between b and c, and so by the difference between the degrees of counterparthood of b to x and of c to x; which latter degrees, of course, will also reflect the *proportion* of parts of b and c which are counterparts to some positive degree. To arrive at a reasonable and consistent assignment of degrees of counterparthood overall is evidently quite complicated, but it does not seem that, on any such assignment, we must obtain a degree of truth for the subformula of (Z) beginning with '$\Box$' which is less than k.

Secondly, to complete the account of fuzzy essences, we want to formulate a crossworld condition with the metalanguage effect of saying that objects which are similar across worlds in design (and anything else deemed to be relevant) and which have most of their parts in common, are counterparts. Again, for a given pair of objects, degree of similarity, which is relevant to the antecedent, will be matched by degree of counterparthood, which is relevant to the consequent, so this principle is also wholly true. Using 'S' for the similarity relation inolved, the principle we want is:

$$\text{(D)}\ \Box_1(\forall x)\Box_2(\forall y)[(\mathsf{A}_1^x\mathsf{A}_2^y(S(x,y))\ \&$$
$$\mathsf{A}_1(MzP(z,x)\colon \mathsf{A}_2P(z,y))\ \&$$
$$\mathsf{A}_2(MzP(z,y)\colon \mathsf{A}_1P(z,x)))\rightarrow x=y].$$

In the counterpart-theoretic metalanguage, (D) becomes the claim that for any world u and object x existing at u, and for any world v and object y existing at v, and for any z which is a counterpart of x at v, if x as it is at u is similar to y as it is at v and most of the parts of x at u are parts of y at v and most of the parts of y at v are parts of x at u, then z = y.[25]

In defusing the paradoxes, we are providing a defence of our view that facts about identity must be intrinsically grounded, but it is evident that the essentialist principles (Z) and (D) are not consequences of this view, since we are no longer employing a framework in which transworld identity features. But the counterpart-theoretic framework may be regarded as a generalization of the standard framework (the necessity of identity is an example of a principle which fails when the standard framework is generalized, but it is a not unexpected

[25] For ease of comprehension, I have not given a counterpart-theoretic interpretation to 'most' here; but, of course, to say that most of a's parts at u are parts of b at v is to say that most of a's parts at u have counterparts at v which are parts of b at v. An unfortunate consequence of Goguen's clause Imp is that if x has counterparts to low degree which also exist at v then (D) will be wholly false, since one of these can be chosen for z while the thing in v with most of x's parts at u can be chosen for y. However, on the revised version of Imp (see notes 23 and 24 above) (D) would then be almost wholly true. We may also prefer to decree that similarity in design and sharing only a few parts is not sufficient for any positive degree of counterparthood.

consequence of a generalization that principles which once held do so no longer) and so we should expect (D) and (Z) to be consequences of a generalized version of the thesis about identity. The generalization is just that transworld heirlines must be intrinsically grounded, or, in Adams's terminology, that facts about realization of thisnesses must be intrinsically grounded. Clearly, our resolution of the paradoxes is in conformity with this thesis, since our whole effort has been towards providing a solution which allows the counterpart relation to hold between objects to just the degree which is demanded by the relevant respects of and degrees of similarity between those objects.

It is left to the reader to establish, by arguments analogous to those used for the oak tree and its acorn, that this requirement of intrinsic grounding cannot reasonably be met by someone who denies (Z) and (D). For these arguments to go through, the branching conception has to be maintained for counterpart theory, but this involves no more than a manageable complication. For instance, in the paradoxical situations we were concerned to model, there was no problem about the transworld heirlines of artefact parts; the difficulty was with what to say about parts put together in different ways or in different groups. Thus the most accurate model reflecting the structure of the situation would be one in which all the worlds have a common initial segment, and the *same* object occurs in the domains of distinct worlds if it exists at a time in those worlds before they branch. It is possible to work out a detailed semantics for a modal language in which the worlds are ascribed further inner structure, i.e. a sequence of times and a domain of existents at those times, although, in the simplest version, it would not be possible to accommodate intuitions about tolerance in *intra-world* persistence conditions. However, the counterpart relation itself would only hold to a degree intermediate between 1 and 0 between objects existing after their worlds have branched; and if it held to maximum degree between two such objects, it would be feasible to allow another exception to the worldbound nature of individuals in counterpart theory, for we could treat identity as a sub-relation of counterparthood, the relation of counterparthood to maximum degree.

The kind of generalization in the modal cases forced by the tolerance principle has its temporal analogue, since the possibility of manipulation of the parts of an artefact through time can give rise to branching, and puzzles about personal identity turn on similar phenomena.[26] For

[26] See the case of Methuselah in Parfit [1971]. Those who think that, in temporal cases involving ordinary material artefacts where there is branching,

instance, if certain sorts of psychological continuity and connectedness are held to be *sufficient* for personal identity, then Oldman would seem to be identical both to Newman-1 and to Newman-2, who are themselves distinct. But here personal identity involves transtemporal identity, and the moral of the paradoxical conclusions is just that when criteria said to be sufficient for identity admit of degree, an appropriate semantics for a language, in this case a tensed language, will employ a relation of degree for the interpretation of sentences which are *de re* with respect to the relevant operators, in this case, tense operators. In tensed discourse, the sentence 'Oldman will be two people' will have some degree of truth; indeed, if each Newman is a counterpart of Oldman to maximum degree, which in this example is a not unreasonable assignment, then this sentence will be wholly true, for there will be distinct, contemporaneous counterparts of Oldman. It may be thought that it is objectionable to extend the counterpart-theoretic apparatus to the temporal case, since it requires an ontology of 'time-bound' individuals, i.e. an ontology of instantaneous things rather than continuants; so the extension apparently implies that there are 'really' no such things as continuants, only time-slices which make up continuants as determined by the counterpart relation. But whatever this might mean, it is no consequence of our view, since we do not take the meta-language sentences of counterpart theory to have the literal meaning which they apparently have; that is, their true meaning is fixed by reverse-translation into the object language. As in the case of the standard framework, the model-theoretic apparatus associated with counterpart theory is simply a convenient device for fixing which arguments are valid, the datum that the paradoxical arguments are invalid being given in advance, by pretheoretic intuition about their object-language formalizations.

Finally, it may be asked why, having set up the counterpart-theoretic

the line of continuity is always the one which traces identity, should consider the following entertaining example, due to David Kaplan. Suppose a museum (in California) hires a philosopher to go to Greece, obtain the Ship of Theseus, dismantle it, pack it and dispatch it back to the museum. Suppose also that the philosopher follows these instructions, but that as he removes each plank from the ship he replaces it with a new plank with the same shape, so that when the original planks are all crated he still has a ship in dry dock. The museum receives the planks, reassembles the ship and is about to exhibit it when it receives a phone call from the philosopher, who announces that *he* has the real ship, and demands large sums of money for keeping quiet about the fraud the museum is about to perpetrate. Assuming the museum director is not an especially dogmatic continuity theorist, will he pay up?

generalization of the standard framework, we do not apply it to entities of every category whose members come from other entities in some suitable sense. A short reply to this is that for e.g. sets and the products of biological growth, there is no tolerance principle analogous to (T). But why is there no such principle for, say, sets, when application of our apparatus blocks the drawing of the unwanted conclusion that a given set could have had completely different members? In connection with *transtemporal* identity, it is not too hard to see why there should be a difference between sets and artefacts, for an artefact is a physical thing which traces a spatio-temporal route, and smallish slowish changes in parts or design are consistent with preservation of enough functional organization relative to the normal state of the artefact for the route to be regarded as that of a single object. With sets, nothing corresponds to a route's having the same structure as it would have had without replacement of part or change in design (these remarks are formulated to allow for dismantling and reassembly). Thus, within a world, there is nothing more to the set than its members, but this is not true of artefacts. There is a comparable difference between the oak tree and a slime mould, since the identity of the oak tree's propagule acorn is not something which can change as time passes. These contrasts are perhaps less marked when we consider modality rather than time, but there is some plausibility in the thought that our conception of the thisness of an individual is fixed by our conception of how it persists through time, equivalently, by the content we ascribe to *de re* tensed sentences about it, and in grasping *de re* modal sentences we simply project that part of the content which embodies our conception of the thisness of the individual to the modal case. Thus, if there is no tolerance in the transtemporal heirline of an entity, our conception of its thisness will not permit tolerance in its transworld heirline.

8

Substances, Properties, and Events

1. *Substances as things*

IN PREVIOUS chapters, we have been concerned with the theory of individual essence for categories of things which we are quite accustomed to thinking of as objects, things whose entitative status is not controversial. The view expounded in those chapters is that individual essences of certain sorts are required by the nature of the identity relation (or some analogue of it), a relation which holds between objects. Thus, to the extent that a certain conception of object seems doubtful, we should expect to find a corresponding weakness in the case for ascribing essences to the putative objects falling under that conception. In this chapter, we will consider three alleged kinds of object, substances, properties, and events, and examine the justification for some essentialist theses about them which others have advanced. This choice of subject-matter is not arbitrary, for we employ modal operators in discourse apparently about such entities, and it is of interest to see how the understanding of modality which we have been developing applies to entities whose nature is somewhat opaque by comparison with material objects and sets. We choose the particular topics of substances, properties and events because there is already a sophisticated literature on them, providing a range of views with which those to be presented here may be contrasted.

Apart from Kripke's ideas about origin, probably the most widely-discussed variety of essentialism is one due to Hilary Putnam, which he motivates by an argument quite unlike anything we have come across so far.[1] This argument might be called the Doppelgänger Argument,

[1] See Putnam [1978, pp. 62–5]. It would be misleading to suggest that Putnam's main concern is to defend essentialist theses about substances; in fact, he is more interested in defending semantic theses about substance terms. For further discussion, see Salmon [1981, Chs. 4 and 5].

and is used by Putnam to attribute essential properties to natural kinds of things, such as species or material substances; for instance, Putnam holds that it is essential to water to be H_2O and that it is essential to tigers to be mammals.[2] So we begin our discussion by first expounding and then evaluating the Doppelgänger Argument.

The argument invites us to perform a thought experiment. We are to imagine a place called Twin Earth, where everything appears to be as it is on Earth; upon investigation, however, some differences manifest themselves, and the essentialist claims are motived by intuitions about how we would revise judgements based on superficial appearance, when the results of such investigations are made known to us. Thus, suppose, as explorers, we set out from Earth and arrive on Twin Earth, where we quench our thirst with the stuff which flows in Twin Earth's rivers and streams and falls from the clouds in Twin Earth's sky, and arm ourselves for protection against the carnivorous large orange and black striped cat-like creatures which hunt in Twin Earth's jungles. So we might report back to Earth that there is an abundant supply of water on Twin Earth, and that on Twin Earth the tiger is not yet an endangered species.

Then our scientists find out that what we have been drinking does not have the same molecular structure as water on Earth; instead of being H_2O, it has some complicated structure XYZ. And after killing one of them we discover that the creatures from which we are defending ourselves are not mammals, like tigers on Earth, but rather, very complex machines built so that they look like tigers and programmed so that they behave like tigers. At this point, Putnam plausibly claims, we would cease to think of what quenches our thirst on Twin Earth as water, and of the carnivorous predators as tigers. When speaking loosely, we might refer to 'Twin Earth water' and 'Twin Earth tigers', but we would agree that Twin Earth water is not the same substance as water, and that the android tiger is not the same species as the tiger. These attitudes, Putnam concludes, betray the view that it is essential to water to be H_2O and essential to tigers to be mammals.

The most straightforward interpretation of Putnam's argument is that it concerns the transworld identity conditions of such entities as substances and species: the transworld heirline of a substance is to be fixed by considerations of molecular structure, or better, abstracting

[2] We will speak throughout of its being essential to a substance to have such-and-such a property although Putnam writes instead that it is necessary. For the purposes of our discussion, the difference is immaterial.

from current physical theory, by considerations of fundamental physical nature; while the transworld heirline of a species is to be fixed by consideration either of relative position in taxonomic or evolutionary trees themselves satisfying some kind of crossworld similarity constraint, or by considerations of genetic and physico-chemical crossworld similarity.[3] Understanding Putnam's argument in this way helps to dispel some kinds of puzzlement its conclusion is apt to produce. For instance, the features on the basis of which the word 'water' is ordinarily applied on Earth are duplicated on Twin Earth; so it may be asked how we can withhold it there without changing the meaning of the word. But on the proposed interpretation of Putnam, we can explain what is wrong with the idea that the meaning of the word must change by an analogy from the theory of reference to material objects. Suppose the explorers have left their friend Smith behind on Earth; however, after a while on Twin Earth, they come across an individual exactly like Smith, that is, an individual who possesses all the features which prompt application of the name 'Smith' on Earth (compare Twin Earth water and 'water'). We can suppose that this individual on Twin Earth even appears to recognize the explorers as old friends, and so they believe that he is Smith, and has transported himself to Twin Earth in some other craft. Then it transpires that this individual too is an android.

In this case, it is indisputable that, despite the great surface similarity between the android and the explorers' friend, these are *different* individuals; it is just wrong to say that the Twin Earth android is Smith, or, more exactly, it is wrong for anyone, such as one of the explorers, in whose language 'Smith' is a name of the Earthman. Similarly, anyone who speaks a language in which 'water' is a name of the substance found on Earth in rivers, streams, etc., will confuse distinct individuals (substances) if he applies 'water' to the water-like substance found on Twin Earth. Exactly parallel remarks may be made about species. It may be true that our ordinary words for animals are less successful in

[3] It is important to note that Putnam is not making an essentialist claim about *members* of the species tiger or about particular *samples* of water. If we treat 'water' and 'H_2O' as predicates of quantities of stuff, then Putnam's claim about water may be regimented:

(i) $\Box(\forall x)(W(x) \rightarrow H_2O(x))$.

The essentialist thesis about individual quantities of stuff with which (i) should not be confused is:

(ii) $\Box(\forall x)\Box(W(x) \rightarrow \Box(E(x) \rightarrow H_2O(x)))$.

marking out natural groupings than are our substance words, but this is irrelevant to the underlying point Putnam wants to make about the transworld identity conditions of such groups.[4] Of course, there are disanalogies between substances and persons as well as analogies; for instance, Smith cannot be on both planets at once, while water can be, but this does not undermine our demonstration that superficial similarity of substances in the two places is hardly sufficient to justify application of the same substance word.

Despite the plausibility of the claims Putnam makes about what we should say about Twin Earth, there is a major doubt about whether the essentialist conclusions he wishes to draw really follow. The problem is that an essentialist thesis about substances should be backed by an argument which speaks of the identity conditions of substances from possible world to possible world, but the story Putnam tells concerns merely a journey within the physical space of a single possible world: the explorers go from planet to planet only. Thus the moral confirmed by the story itself is just that the same substance is present in different *places* (within a world) only if its instances at those places all have the same fundamental physical nature. And this moral is compatible with the existence of a possible world all of whose samples of water have molecular structure XYZ. Putnam does not seem concerned by this gap between his story and his conclusion. He writes:

> Suppose, now, that I discover the microstructure of water . . . that water is H_2O. At this point, I will be able to say that the stuff on Twin Earth that I earlier mistook for water is not really water. *In the same way*, if you describe, not another planet in the actual universe, but another possible universe in which there is stuff with the chemical formula XYZ which passes the 'operational test' for water, we shall have to say that that stuff isn't water, but merely XYZ.[5]

In this extract, we emphasize the words 'In the same way', since this phrase appears merely to mark a lacuna where an argument is needed, rather than to signal an unproblematic extension of the given considerations to an analogous case.

Someone might try to fill the lacuna by appealing to the parallel with ordinary objects, like persons, where most, perhaps all, of the features of a person which prompt our recognition of him are contingent

[4] In [1981] Dupré assembles much evidence that our ordinary words for kinds of animals and plants do not pick out the natural kinds recognized by science. But this is irrelevant to essentialist theses about kinds, which would just have to be expressed using technical terms for species.

[5] Putnam [1978, p. 70].

features, while a feature we do not have access to in ordinary circumstances, his origin, is essential. It might be said that just as the zygote is to the organism, so physical nature is to the substance. But this cannot just be said. We need to see the argument that someone who denies such a thesis is committed to bare truths or extrinsicness in connection with transworld identity, and also an argument that even for such esoteric entities as substances, bare truth or extrinsicness should not be tolerated. So it looks as if the Doppelgänger Argument is a complete red herring, and that a successful argument for Putnam's conclusion about the essential properties of substances would have to be pursued by the same methods as were used for sets, organisms, and artefacts in previous sections.

2. *Identical substances are necessarily identical*

There are other arguments in the literature for the essentiality of fundamental physical nature which do not, or need not, rely on Doppelgänger-type thought experiments; these other arguments might be termed 'semantic'. The simplest is due to Kripke, according to whom 'Heat is the motion of molecules', 'Water is H_2O', and other examples, 'are all in some sense of "identity statement", identity statements'.[6] Then if 'Water is H_2O' is a true identity stagement, 'Necessarily, water is H_2O' is a truth delivered by the necessity of identity. And even if we move to a counterpart-theoretic framework, in which the necessity of identity is not valid, the evidence of the Twin Earth thought experiment suggests that this will not lead to the kind of counterexample to the essentialist claim which the sceptic wants. However, three points may be made about Kripke's argument, which appear considerably to undermine it.

(i) If 'Water is H_2O' is an identity sentence, then 'water' and 'H_2O' must be terms picking out entities. So far, we have referred to these entities as substances, but exactly what kind of entity is the substance water (or, for that matter, the species tiger)? Certainly not the Goodmanian sum of actual samples of water, since it is the individual essence of that sum of water to be composed of exactly the water it is in fact composed of, while it seems true to say that there could have been more or less water than there actually is. So it seems that the substance (species) must be an abstract entity of some sort and, if this is so, then it is no longer clear that 'Water is H_2O' is even true. For, on this

[6] Kripke [1971, p. 143].

construal, all that science will have discovered is that actual samples of water have the structure H_2O, and it is controversial that actual co-extensiveness is sufficient for identity of the abstract objects, which are perhaps the properties of being water and being H_2O; whether scientists should be said to have discovered an identity of properties or merely an identity of the sets of things which actually have those properties is a philosophical question on which a theory of the essences of substances should remain neutral, if possible; but Kripke's argument presumes a particular answer to that question.

(ii) Another presumption of the argument is that it is definitely correct to say that ice and steam are forms of water, water in the solid and gaseous states respectively, rather than that ice, water and steam are three forms of H_2O. For if we say that latter, then 'Water is H_2O' is not an identity sentence (if it is true) but, rather, is elliptical for 'Water is H_2O in the liquid state'. Again, it is unclear why a theory of essences should have to rule that one form of speech is the right one, and that we do not have the 'is' of predication in 'Water is H_2O'.

(iii) If 'Water is H_2O' is an identity sentence, then its necessitation follows only if the terms flanking the 'is' of identity are rigid designators. However, 'H_2O' seems to have semantically significant structure; perhaps it abbreviates a definite description such as 'the substance pure samples of which have molecules composed of two atoms of hydrogen and one of oxygen in such-and-such a configuration'. Is this description a rigid designator? That is, does it denote the same substance at every possible world, or even just at every world where it denotes at all? A sceptic about the necessity of water's being H_2O can afford for the sake of the argument to concede that 'hydrogen' and 'oxygen' are rigid designators: he may still ask why it should be thought that putting two hydrogen atoms and one oxygen atom together in that configuration in *any* world always yields a molecule of the *same* substance. In other words, on the hypothesis that 'H_2O' abbreviates such a definite description, assuming that it is a rigid designator is equivalent to assuming that what it is to be the same substance in different worlds is to have the same molecular structure, or fundamental physical nature, in those worlds. Obviously, this is the doctrine which is to be established, so a simple argument from the necessity of identity is quite circular.[7]

[7] Similar points are made in Salmon [1981, ch. 6]. In Almog [1981] it is argued that the sense of 'water' has an indexical element: what 'water' refers to is determined by a feature of the context in which the word is introduced into our language; it refers to the substance which, in that context, had certain phenomenal features (the context is a composite of place, time and possible

3. *Crossworld equivalence relations*

We can avoid the difficulty of specifying the entity to which 'water' refers if we treat the word as a predicate. Putnam emphasizes the fact that 'water' is introduced by ostension of samples, and we can regard such ostensive sentences as attributing a property, that of being water, to objects, the quantities of stuff which are the samples. Suppose two such quantities are both water; then although we might say that they are two quantities of the same substance, this need not commit us to an ontology of substances, for the statement that this stuff is a sample of the same substance as that stuff can be regimented as an atomic sentence of the form Rtt', to the effect that this sample is *cosubstantial* with that one. This move undercuts any argument for its being essential to water to be H_2O which turns on treating substances as entities and applying our earlier considerations concerning transworld identity to them (any defence of the essentialist thesis for which it was crucial

world). But Almog is unable to tell a consistent story from this premiss, for he wants to say *both* that 'water' is directly referential, that is, it introduces its extension into propositions expressed using it, where the extension at a world is (p. 352) 'an appropriate body of liquid' ('The only sensible way is to acknowledge that . . . [substance terms] . . . bring into the propositions the extensions themselves' (p. 354)) *and* that these terms bring their contents into the propositions, where the content is an intension, i.e. a function from worlds to bodies of liquid (p. 353). And he also professes metaphysical agnosticism about what the reference of 'water' is (pp. 354–5). Part of the trouble here stems from a problem about the underlying theory of indexicals, due to Kaplan [1977], in which the content of an indexical in a context of use is at once an intension, a function from worlds to extensions (a constant function) and the extension of the indexical in that context. One must make up one's mind which of the two it is to be. Moreover, the spirit of Kaplan's theory demands that the content be the extension, but this is not a possible choice for 'water' if it is to be true that there could have been more or less water than there actually is. But quite independently of how these details are to be worked out, it is not very plausible that substance terms are indexicals, i.e., words whose semantic values or contents vary from context to context. It may seem as if this is right for the Twin Earth case, but only when the redundant extra feature of Twin-Earth-English speakers is added to the example; then their use of 'water' and our use seems like your use of 'I' and mine. Without this feature, the example is just like one in which real gold is found in one part of this planet and fool's gold in another; someone who uses 'gold' for both substances is using the word *ambiguously*, and if there are speakers in the area where fool's gold is to be found who call it 'gold', we should say that *either* they speak our language and use the word ambiguously, or, more likely (assuming they have no contact with us of the kind which makes it plausible to regard them as members of our linguistic community) that they speak another language in which, by sheer coincidence, they use the same word as we do for a substance which looks rather like the substance to which we apply that word. It would be quite unmotivated to speak of indexicality in this case, which is not significantly different from Putnam's.

that substances be treated as entities would *ipso facto* be unsatisfactory). Our point about the Twin Earth thought experiment may then be expressed by the objection that although the experiment establishes that the cross-spatial and cross-temporal (*intraworld*) application conditions of the relation of being cosubstantial with involve the fundamental physical nature of the samples, it does not follow that the same is true of the crossworld application conditions.

However, to say that it does not follow prompts the more general question of how the intraworld application conditions of a two-place relation in modal language are affected if that language is equipped with the doubly-indexed 'actually' operators so that the relation can be used to make crossworld comparisons. If 'C(a, b)' means that a is a sample of the same substance as b then

$$\Diamond_1 \mathsf{A}^a \mathsf{A}_1^b C(a, b)$$

makes such a comparison. In possible worlds terms, it says that there is some w such that a as it is in the actual world is cosubstantial with b as it is in w. For 'b' we could then put some descriptive name 'β' of a counterfactual sample of a substance much like water except for having the structure XYZ. With this apparatus, we might then try to fill in the lacuna in Putnam's argument as follows. The Doppelgänger arguments show that, within a single possible world, application of the relation 'is cosubstantial with' is constrained by facts about fundamental physical nature. To ignore this constraint would be to change the meaning of that relation. But adding extra modal operators to modal language cannot affect the sense of a relation-symbol already in the language, and so we cannot ignore the constraint in crossworld applications of the relation. Hence

$$\Diamond_1 \mathsf{A}^a \mathsf{A}_1^\beta C(a, \beta)$$

must be false.

To evaluate this argument, we should consider some other cases of equivalence relations. First, take the relation 'is sameshaped with'. It is clear that the intraworld and crossworld application conditions of this relation are the same. A related fact is that, in cases where the relation holds, there is a level at which it does not make sense to ask in virtue of what it holds: such cases afford examples of bare truths. Even if we agree to say that two objects are sameshaped in virtue of their both being square, we cannot say what the first's being square, or the second's being square, consists in: it would simply double-count the same fact to speak here of sizes of angles and lengths of sides, in the way that it would double-count the same fact to say that a man's being a bachelor *consists in* his being male and not yet married (so the relationship of analytic equivalence is not what we intend by 'consisting in'). Contrast this with 'is of the same nationality as'. One could say that a and b are of the same nationality in virtue of their both being citizens of the United Kingdom, just as two shapes may be sameshaped in virtue of their both being square. But, in this case, one can also say in virtue of what fact a, or b, is a citizen of the UK. This may be in virtue of having been born there, or in virtue of having undergone a naturalization process, or in virtue of the obtaining of any other condition which the UK authorities happen to deem to be sufficient for citizenship. Moreover, in making a crossworld application of 'is of the same nationality as', as in 'if b had been thus-and-so, then b would have had the same nationality as a actually has', we may imagine a world in which some condition suffices for being British which does not actually suffice, and we can do this without changing the meaning of the relation.

How does 'is cosubstantial with' compare to these cases? As before, we can say that a and b are cosubstantial in virtue of both being samples of water, but it also seems correct to say that, for the intraworld case, each is a sample of water in virtue of its having a certain physical nature. We expect the superficial and easily detectable differences between pieces of stuff to reflect fundamental differences which explain the superficial ones, and it is the fundamental differences which have the final say in classification; so someone who refuses to classify samples in this way may fairly be said not to understand what a substance is. However, before concluding that 'is cosubstantial with' is like 'is the same shape as', and that the 'same physical nature' criterion should be projected across worlds, we should ask why this criterion is appropriate for intraworld applications. The conception of substance that goes along with the criterion is made for our conception of the physical universe as a law-governed causal system, for one consequence of claiming that a and b are samples of the same substance is that, *ceteris paribus*, they may be interchanged in a fixed type of environment without affecting the outcome of processes unfolding in that environment; it is thus that experimental results confirm or disconfirm an application of 'cosubstantial'. This consequence derives from our understanding of the causal powers of underlying physical natures—there is a common *explanans* in causal explanations of the behaviour of two samples of the same substance in the same conditions—together with the reliability of the laws governing causal connections.

If these remarks about the intraworld application of 'is cosubstantial with' are correct, it follows that in making a crossworld application of this relation the rationale for the criterion of intraworld application is lost; for, of course, distinct possible worlds cannot constitute a single causal system. Here there is an analogy with 'is the same F as', where F is a sortal for material objects, and a contrast with 'is the same set as'; for none of the standard criteria for intraworld identity of F's, e.g. spatio-temporal continuity, can be applied across worlds, while there is no problem in applying 'has the same members as' across worlds.

This conclusion may seem inconsistent with our earlier advocacy of the branching conception of possible worlds, a conception which effects a reduction of some instances of crossworld identity to intraworld identity. When we are dealing with samples of the same substance there seems to be nothing corresponding to the possibility of an object's coming into existence in a world after that world has branched from the actual world, so it might be suggested that the branching conception reduces all applications of 'is cosubstantial with' to intraworld applications. But this is incorrect, since a world might branch from the actual world before there is any sample of the relevant substance in the actual world. Furthermore, the branching conception does not assert that every possible world branches off from the actual world, only that every world with an object in common with the actual world branches from it; there are other possible worlds (in some sense 'qualitative') which do not represent possibilities for any actual object, or for any non-actual object whose identity can be fixed by predicates and names of actuals (such as the artefact β from the Four Worlds Paradox), and it cannot just be *assumed* that 'is cosubstantial with' has no application between such a world and the actual one.

Hence we have failed to discover any compelling ground for holding that the relation of cosubstantiality gives rise to essentialist truths of the kind it has been credited with the power to generate; the possibility remains open that, like 'is of the same nationality', a particular substance-predicate may be applied in distinct worlds on the basis of distinct criteria. In the case of 'same nationality', what justifies us in claiming that b is British in a world where there are new ways of coming to be British is that the nation of which b is a citizen in that world is indeed the UK, and that 'has British citizenship' is applicable to him in virtue of the consequences of the status he has with respect to the UK in that world—he has the same rights and privileges as UK citizens actually have, for instance. One reason why the necessity of water's being H_2O seems plausible, perhaps, is that it is not easy to think of a counterexample in which there are features corresponding to these in the citizenship case.[8] But that is hardly an argument *for* the essentialist claim, about which this discussion justifies some agnosticism.[9]

[8] Unlike the nationality case, there is a problem in agreeing upon what would count as the *same* consequences. But suppose that matter is non-atomic and yet some substance exhibits all the more obvious characteristics of water, although its composition is not proportionally two to one of anything. There may be no compelling reason to apply 'water' to portions of such stuff, but there does not seem to be any compelling reason not to, either.

[9] How do these considerations apply to the question of the essential properties of species? In [1976] McGinn argues that the necessity of origin applies to species and endorses Dummett's remark that 'even if creatures exactly like men arose from dragon's teeth, they would not be men, because not children of Adam.' (Dummett [1973, p. 144].) In our terminology, the claim is that the crossworld extension of 'is a member of the same species as' contains the 4-tuple ⟨a, u, b, v⟩ only if the species of a's ancestors in u are the same as the species of b's ancestors in v. How strong is the case for this claim? The *intra*world application conditions of 'is cospecific with' do not refer to ancestry (although this is relevant to the generic classification of species) and so there is even less support for the essentialist thesis here than there was for the one about substances, where intraworld application conditions are answerable to physical nature. The main criterion for intraworld application of 'is of the same species as' is that of reproductive behaviour: in a population of fauna within a particular geographical region, the boundaries of the species represent barriers across which mating, or at least the production of viable offspring, does not naturally take place. Evidently, this criterion is straightforwardly applied only to organisms which reproduce sexually, and only to populations in the same geological period and same geographical region. However, when these conditions are satisfied, the decision whether or not to discern one species or two in a given population is still answerable to our view of the correct shape of the evolutionary tree: it is just that reproductive isolation in such circumstances is the best evidence for what that shape should be. When the conditions are not satisfied, there is an element of arbitrariness in taxonomic classification (see Maynard Smith [1975, pp. 209 ff.]). Where populations in different periods are concerned, taxonomists tend to classify using similarity measures derived from decisions taken in cases where the three conditions are satisfied. However, if fossils suggested an earlier species very similar to a current one, but there was good reason to believe that the earlier species had a very different descent, a distinction between the species would be maintained. Again, however, this crosstemporal constraint is underpinned by the idea of a single evolutionary history encompassing all life on Earth, a conception which has no crossworld application. So if Dummett's creatures were exactly like men, this kind of reason for not counting them as members of the species *homo sapiens* lapses. Perhaps counting such possible creatures as of the same species as actual men implies commitment to counterfactuals about mating behaviour and the viability of offspring, but there is no reason why such counterfactuals could not be true.

4. *Properties*

It is possible to think of properties as entities of a certain sort; for instance, if the substance water and the property of being water are the same thing, then someone who construes the discussion above to have been about the transworld identity conditions of substances may think of it as a special case of the more general problem of the transworld identity conditions of properties. This problem would also have arisen earlier if instead of formalizing 'my car could have been the same colour as yours actually is', using the doubly-indexed actually operators (see §5 of Chapter Four), we had regimented it as:

$$(1)\quad (\exists x)(x \text{ is a colour } \& \text{ Applies to}(x, \text{your car}) \; \& \; \Diamond(\text{Applies to}(x, \text{my car}))).$$

If (1) is intelligible, and if transworld identity must be intrinsically grounded for all categories of entities, then it must make sense to ask in virtue of what the colour property which applies to your car in the actual world is the same property as the colour property which applies to my car in a world which verifies (1); and the same can be said for any property of a given type, a type which can be expressed by a predicate (other than a sortal for ordinary things) which can replace 'colour' in our sentence, or one like it, e.g. 'size', 'shape', 'make', 'design', 'age' and so on.

Our view is that such transworld identities of properties reduce to the holding of other transworld relations between ordinary individuals. Thus to say that a certain shape property in u is the same as a certain shape property in v is to say that the things in u to which the first shape property applies are sameshaped with the things in v to which the second shape property applies; so the reduction is effected by a cross-world equivalence relation which holds between things of the kind to which the property applies within a world. But as our discussion of substances reveals, it is not always straightforward to decide what the crossworld criteria of application of the equivalence relation should be; for instance, consider the problem of applying 'is sameshaped with' across worlds with radically different physico-geometric properties. However, the intraworld criteria provide guidance up to a point, for it seems correct to say that if there is nothing in virtue of which the relation holds within a world, that is, if its holding in a particular case is a bare fact, then its holding across worlds in a particular case will also be a bare fact. Consider, for instance, the example of length. Someone might suggest that, within a world, two objects are of the same length iff, were they laid alongside one another, the corner points of the adjacent edges would be in contact. But apart from the fact that in the present context it is unhelpful to try to ground an apparently non-modal fact in a modal fact (the one expressed by the counterfactual), such counterfactual analyses do not work unless the concept they are analysing is dispositional; for it may well be that although two objects are of the same length, circumstances are such that if either were or had been moved, its length would change or would have changed. Similar objections can always be found to proposals of this kind, which shows that being of a given length is not the same as being disposed to affect measurement instruments in a certain way, nor indeed to behave in any other particular fashion.

However, not all properties are like length properties. Some are explicitly dispositional, like the property of fragility, which is the disposition to fragment if subject to an impact of a certain force.[10] But there is no mystery about the crossworld application conditions of 'is disposed to fragment like'; what is required is that the same counterfactuals be true of each object at its own world. It is *not* required, nor is it sufficient, that the objects have the same physical nature; for a physical nature which realizes the dispositional property of fragility at one world may not do so at another; and the same counterfactuals might be true at different worlds of objects with different physical natures across those worlds.

Another interesting class of properties are those which Locke called 'secondary qualities', which he identified with powers in objects to affect us in certain ways.[11] The property of being yellow, for example,

[10] Fragility is not just the disposition to fragment when involved in the initial stages of a process which in normal circumstances would result in an impact being suffered; rather, the impact must occur. But even then the outcome of the impact is not all that matters. Suppose that an object would shatter if subject to a certain impact but for God's intention to disrupt the causal efficacy of the impact so that no shattering occurs although the impact does. Then even if God intends to do this every time, such an object should still be said to be fragile, since the normal process which leads from impact to shattering is being subject to interference from outside.

[11] See Locke's *Essay Concerning Human Understanding*, 2. vii., 14–17. Unfortunately, Locke says the same thing about primary qualities (loc. cit., 8), which are supposed to differ from secondary qualities in virtue of producing ideas in our minds which resemble the primary qualities themselves. But if primary qualities are also dispositions, it is hard to see what 'resemble' could mean here. We may interpret Locke charitably, as intending the dispositional/non-dispositional contrast.

could be identified with the disposition to produce in normal observers and in standard observational circumstances visual experiences with a certain characteristic phenomenal quality, what one might call 'phenomenal yellowness'.[12] Thus applications of 'is of the same colour as' within the actual world are relative to a conception of normality which may not be appropriate for another world in which objects are nevertheless coloured; what matters are *actual* standards of normality. A more radical account is that e.g. 'yellow' *means* 'produces phenomenally yellow sensations in normal observers in standard circumstances', so that for an object to be yellow at a world w it is required to affect observers in a certain way *at w*. Kripke has objected to such an account that

> 'yellow' does not *mean* 'tends to produce such and such a sensation'; if we had different neural structures, if atmospheric conditions had been different, if we had been blind, and so on, then yellow objects would have done no such thing.[13]

However, a defender of the radical view could account for the apparent plausibility of this objection by appeal to a scope distinction. What is

[12] See Peacocke [1983a] for a theory of 'phenomenal colour' which elegantly resolves the quandary of the relative primacy of 'F' and 'looks F', for F a colour word. The idea that 'yellow' is so much as coextensive with 'produces experiences with such and such features in normal observers in normal circumstances' has been contested in Averill [1982], where the author points out that a surface made up of tiny red and green dots looks yellow to normal observers under normal lighting conditions, but is really red and green; that a black and white disk can appear to have red, yellow, and blue bands when it is spinning; and that a small green patch on a large yellow surface looks blue to normal observers. This may suggest that there is a notion of what the colour of an object *really* is which our dispositional account does not explain, but other possible morals are that there is a non-circular reason for saying that, in these cases, circumstances are not normal, or, contrary to Averill's implied assumption about the cases, that there is no clear matter of fact about the colours of the objects, because it is unclear which circumstances are normal, and an element of decision, guided by other cases, is involved. This last option appears attractive. For instance, the facts we are faced with in the case of the red and green dotted surface is that it produces sensations with the phenomenal-yellow quality in some conditions and that it produces sensations with phenomenal red and phenomenal green dotwise intermingled in others (close-range observation). Neither type of circumstance has an obvious claim to be regarded as the 'normal' one, but, if we have to choose it is hardly inexplicable that we should prefer the circumstance of close-range observation (similarly for the preference for observation of the disk when stationary and of the small patch when not embedded in a rather special background); in particular, there is no reason to suspect that this choice cannot be justified without presuming upon a fact about what colour the object really is.

[13] Kripke [1972, p. 354]. Kripke appears to hold that a non-circular specification of 'normal circumstances' is impossible (loc. cit.). See the previous note for a reply to this.

true is that, concerning things which *are* yellow, if we had had different neural structures, these things would not have produced phenomenally yellow sensations in us, which means that in *those* circumstances they would not have been yellow. And if we had evolved without the capacity for visual sensations, nothing would have been coloured. So, according to this view, there is no particularly intimate connection between colours and wavelengths of light, other than the contingent connections which scientists have discovered actually to obtain.[14]

5. *Events*

We shall take events to be dated, unrepeatable occurrences occupying definite intervals of time. Clearly, the time of an event contributes to its intraworld identity condition. A very rough and intuitive account can be set out by adding to the time of an event two further components, the objects *involved in* the event, and the types of changes in properties which these objects undergo for the duration of the event; this will be elaborated and qualified in due course. We might involve the location of an event in its intraworld identity condition, but such a component would be redundant, since the same objects cannot be in different places at the same times; so the objects and times fix the places, whereas place and time does not fix object (a statue and the bronze of which it is made are different objects, although they are in the same place). On this intuitive view, then, an event consists in a triple of (i) a set of objects; (ii) types of changes of properties for each object in the set; and (iii) an interval of time. Obviously, not any combination of items falling under (i), (ii), and (iii) constitutes so much as a possible event; while we say that a triple *constitutes* an event, rather

[14] Kripke's view, also that of Davies and Humberstone [1981], is that colour words stand for properties which physicists can mark out in more fundamental terms, presumably terms involving the wavelength of emitted or reflected light. Thus, if we had been different in certain ways so that red objects had looked yellow to us and we had applied 'yellow' to them instead of to the objects which in those circumstances would have looked yellow to us had we not been thus different, we would be picking out with that use of 'yellow' a different property from the property we actually use it to express. The Kripke–Davies–Humberstone view does justice to Locke's idea that colours are *powers* in objects to affect us in certain ways, since that view identifies the colour property with the physical ground of the disposition. The radical view does justice to Locke's idea that, in some sense, colours are not really 'in' the objects themsleves. Here the thought is that primary and secondary qualities differ in that the *identity* of secondary qualities at a world depends upon facts about the make-up and environment of sentient beings at that world.

than is identical to it, to leave it open that one and the same event may be constituted by different triples in different worlds. Finally, it might be preferable to treat types of changes of properties as fixing a type for the event, like sorts for objects; types, like sorts, will be specified by expressions of a special category, the type of an event being, very crudely, the 'basic' change which is exemplified when the event occurs, in a sense of 'basic' yet to be explained. So an event of type T would be an exemplification of T by a set of objects and an interval of time.

In a famous footnote to a discussion of counterfactuals about events, David Lewis gave a clear statement of an intuition about events which appears to show that *de re* locutions are quite unproblematic in connection with them.[15] Lewis was discussing whether or not a certain event, the death of Socrates, would have occurred if things had been different in certain ways, and he points out that to ask whether *that* event would have occurred is not the same thing as asking whether there would have been a unique death of Socrates. For we can imagine that everything is as it actually is up to and through the death of Socrates, and then Socrates is resurrected to die a second time. In such possible circumstances, it is plausible that his first death is the same event as his actual death, but it is not a unique death. Equally, Socrates might have died in a totally different way from the way in which he actually died, and plausibly the event which was his actual death does not occur in such circumstances. Here Lewis is apparently relying on an intuitive conception of *the event itself*, apart from such descriptions of it as 'the death of Socrates'; the event has a thisness under which it can be projected across worlds, a thisness which may either be primitive or else intrinsically grounded. If the latter, then there must be favoured descriptions of the event, which will fasten onto its individual essence.

Opposed to Lewis's view is a position which we will call *de re* scepticism, which says that the conception to which Lewis hopefully appeals simply does not have any content: there is no such thing as *the event itself.* Since we are being neutral on the question of whether or not Lewis's conception is completely analysable, it follows that *de re* scepticism is stronger than the denial of the thesis that events have individual essences. Such a sceptic could make sense of *de re* modal discourse involving quantification over events by employing some relatively stipulative or extrinsically grounded criteria for transworld

[15] Lewis [1973b, footnote 9].

heirlines, a position which would be consonant with a degree of scorn towards the alleged entitative status of events themselves.

To proceed further, our rough and ready sketch of the nature of events needs to be elaborated; fortunately, we can avoid extra work here by simply availing ourselves of the most sophisticated and complete development in the current literature of the intuitive ideas about events employed above, due to L. B. Lombard.[16] The central idea of Lombard's theory of the nature of events is that of a *quality space*; events are movements of objects at times in such spaces. A quality space is a space of static properties, properties possession of which does not imply change (compare being five feet in height with growing). Quality spaces are *closed* under certain kinds of physically possible changes of static properties, and the individuation of quality spaces determines the individuation of events. Let us confine ourselves, as Lombard does, to events which are changes in a single physical object, such as the death of Socrates, as opposed to, say, a mass suicide. So, for instance, if an object changes shape during an interval of time while it is also changing location, we will say that two events are occurring, since the object is moving simultaneously in two different quality spaces. Static shape qualities and static location properties are assigned to different quality spaces since changes of shape and changes of position are unconnected, in the sense that an object's position can change without its shape changing, and conversely. By the closure condition, *all* shape properties are in one space and all location properties in another.

An object may change in virtue of having a part which changes; in Lombard's view, there is exactly one event occurring in such situations, which involves both the larger object and its part. An object is then said to be *minimally* involved in an event e iff it is the smallest object a change in which is identical to e. When a change in an object is not to be construed as a change in other objects out of which the first is composed, Lombard says that the object is *atomic*; an atomic quality space is a quality space with only properties which it is physically possible that an atomic object has, and an *atomic event* is a movement of a single atomic object in a single atomic quality space. This gives us a basis for a classification of events; each event is either an atomic event, a temporal sequence of atomic events, an event composed of simultaneous atomic events, or a temporal sequence of such composite events. From these materials, Lombard derives the notion of a *canonical*

[16] See Lombard [1979, 1981, 1982].

description of an event. An atomic event's canonical description is a singular term of the form

$$[s, \phi, t]$$

where s is a name of the subject of the event, t is the time interval the event occupies, and ϕ is an expression for the gaining and losing of the static properties gained and lost by s in its movement in the quality space; the singular term schema might be read 'the ϕ-ing of s during t'. A non-atomic event's canonical description is constructed out of the descriptions of the event's atomic constituents, so as to describe how those events are bound together into the non-atomic event. So in Lewis's case, the death of Socrates is a non-atomic event whose canonical description will involve an account of the physical effect on his body of the poison he swallowed.

6. *Lombard's essentialism*

Canonical descriptions seem to be the favoured descriptions the essentialist is looking for under which an event is projected into counterfactual situations, or pronounced to be absent from such situations; not that those who use *de re* locutions in connection with events must be able to come up with such a description, but the availability of it legitimizes the use. If canonical descriptions are favoured in this sense, then it must be essential to events to occur when they do and to have the minimal subject which in fact they do have. Lombard argues for both theses. *Prima facie*, it may seem obvious that even the minimally involved subject of an event is accidental to it, but this appearance rests upon a scope confusion. Since β might have been the particle with the positive charge instead of α, it is possible that the impinging of the particle with the positive charge on the photographic plate involves β as minimal subject rather than α. But the obvious truth here bestows no plausibility at all on the claim that the actual impinging of the charged particle might have had β as minimal subject. However, Lombard constructs an ingenious case for the anti-essentialist about minimal subject (or for the outright *de re* sceptic) in which there is no scope confusion, before trying to show that, even in this hard case intuition is on the side of the essentialist.[17]

Suppose that in u, the Ship of Theseus is constructed from planks $p_1 \ldots p_n$ and that in u the Ship of Theseus sinks in a certain storm in

[17] Lombard [1981, pp. 142–5].

a certain place and time. Let v be a world resembling u as much as possible compatible with the following conditions obtaining in v: the Ship of Theseus undergoes a gradual replanking process until it comes to be made of planks $q_1 \ldots q_n$, while planks $p_1 \ldots p_n$ are assembled into another ship, the pseudo Ship of Theseus, and on the fateful day in v it is this second ship which sinks. The sceptic may then argue: in u, the sinking of the Ship of Theseus, e_1, is the 'sum' of the sinkings of planks $p_1 \ldots p_n$, which sum we call e_2 (so $e_2 = \text{sum}[e_{p_2} \ldots e_{p_n}]$). It is hard to deny that e_2 occurs in v. But then by transitivity of identity, e_1 occurs in v and is the sinking of the pseudo Ship, since it is this ship's going down which, in v, is identical to the event $\text{sum}[e_{p_1} \ldots e_{p_n}]$.

It would be somewhat wooden to dispute the example on the grounds of its assumptions about transtemporal and transworld identity for ships (questions about the transworld heirlines of events may presume upon the transworld heirlines of the relevant non-events). Instead, Lombard attacks the premiss that e_1 is identical to $\text{sum}[e_{p_1} \ldots e_{p_n}]$, by claiming that e_1 has a modal property which e_2 ($= \text{sum}[e_{p_i}]$) lacks: it is essential to e_2 to have each e_{p_i} as a part, but since the Ship of Theseus could have been made of planks $p_1 \ldots p_{n-1}, q_n$, e_1 might have had e_{q_n} as a part instead of e_{p_n}. Or, more simply, we can just say that e_1 could have occurred without e_{p_n} occurring, while e_2 could not have so occurred. However, like all arguments for distinctions established by possible discernibility, this objection may be met by disputing the underlying transworld identity judgement, in this case that e_1 occurs in some worlds where the Ship of Theseus has q_n in place of p_n. But someone who denies transworld identity between e_1 in u and the sinking of the Ship of Theseus in any world where it has q_n must offend our stricture that transworld distinctness cannot be imputed where there is nothing intrinsic in which the difference consists. For, with respect to some worlds, there will be no difference at all between the sinking of the Ship of Theseus in those worlds and its sinking in u: the sinking will occur in the same manner, at the same time, in the same place, and will involve the same ship *ex hypothesi*. That is, the natures of the events, at the very least, are the same in the different worlds, so there is nothing in which a numerical difference between any pair of them could consist. In this fairly minor way, then, our thesis that facts about identity must be intrinsically grounded is already relevant.

The anti-essentialist may try a different manoeuvre, which Lombard does not consider in this context, which involves putting forward a sufficient condition for transworld event identity in terms of causes, that if e has exactly the causes in u which f has in v, then e and f are the same event (since he is an *anti*-essentialist, he will deny the converse, but the objection could also be made by an essentialist with an alternative account to Lombard's). According to this condition, the following is a counterexample to Lombard's view. Let e be an actual event involving an object x at rest at place p at time t and moving away from p thereafter, the causes of this movement being events which are impacts of other objects, these events contiguous in space and time to e. Since it is contingent that x is at p at t, we can choose a world w with another object y at p at t such that w is as similar as possible to the actual world up to and including t. Then the very same impacts of other objects which caused e in the actual world also occur in w, but in w they bring about a movement of y (a movement of the same type, even if not the same movement). By the objector's sufficient condition for event identity, therefore, y's movement in w *is* e. So e could have minimally involved y instead of x.

The strangeness of this sufficient condition appears to be traceable to an intuition about what the intrinsic features of an event relevant to its crossworld individuation are. In the case of material objects and sets, such intrinsic features were invariably in some sense 'internal' to the thing, either its members, or its constitution and design, or its origin and kind. Although an event's causes are obviously not causally isolated from it, they are not internal to it in a way analogous with the material object case, while the minimal subject clearly is internal. For this reason, the idea that having the same causes is sufficient for crossworld identity of events may be rejected.

Lombard's second essentialist thesis is that the time of an event is essential to it, where by 'time of an event' we mean the interval exactly occupied by it. Lombard argues for this thesis by confronting the sceptic about it with a *reductio.*[18] If an event can occur at different times within different worlds, then if we take a pair of events e_1 and e_2 occurring in a world u which are intrinsically indistinguishable in u except by the time of their occurrence (Lombard uses the example of a marble's rolling twice along the same route at the same speed) there must be worlds in which e_1 and e_2 switch times, so that e_1 occurs at t_2 and e_2 at t_1. From amongst those worlds, choose v as similar as possible to u but for the switched times of occurrence; then it seems that the only substantial difference between u and v will be in the identity of the events which occur at t_1 and t_2 in the two worlds which

[18] Lombard [1982, pp. 9–13].

involve that marble's moving, since the intrinsic indistinguishability of the events implies that the antecedents of one do not have to be intrinsically distinguishable from those of the other. Thus, although Lombard does not put it in this terminology, we see that rejection of essentialism about time leads to ungrounded differences between entities and resulting inadmissible distinctions amongst worlds much like that which we arrived at by the Four Worlds Paradox.[19]

Lombard's strict essentialism about the time of an event could be relaxed by allowing a small amount of change, so that the *same* thunderstorm could have started a little later, but not a day later. Such a modification would require application of the degree- and counterpart-theoretic apparatus to modal discourse about events, but this does not constitute a very great change to Lombard's approach. A more radical worry concerns whether or not Lombard's theory does capture genuine aspects of our concept of event. Crucial to the construction of the theory is Lombard's use of the *de re/de dicto* distinction to reply to certain objections, but while the content of the distinction is well understood in the case of quantification over ordinary objects, a sceptic about essentialism about events may feel that too much is already conceded if a distinction is allowed between its being possible for the death of Socrates to have occurred earlier and the death of Socrates'

[19] Lombard's own view is that what is wrong with taking u and v to be numerically distinct worlds is that it contravenes a supervenience principle for events which he states thus: (E) Possible worlds cannot be alike with respect to the truth and falsity of propositions concerning objects, the properties which those objects have, and the times at which those objects have those properties, and yet be unalike with respect to the truth and falsity of propositions concerning events. Lombard thinks that this principle has all the plausibility of an analogous supervenience principle for sets: (S) Possible worlds cannot be alike with respect to the truth and falsity of propositions concerning the existence of objects, and yet be unalike with respect to the truth and falsity of propositions concerning the existence of sets. Each principle, he says, expresses the doctrine that there is nothing 'hidden' about sets or events, nothing more to know about them once we know their natures (in the case of sets, their members) [1982, pp. 13–14]. But principle (S) does not have this effect, since it is compatible with the same set having different members in different worlds with the same domain of non-sets; furthermore, it entails a restricted version of Set Existence, that if all the members of x at some world u exist at v and v has the same domain of non-sets as u, then x exists at v. But Lombard explicitly rejects the idea that the existence of a set 'follows simply from the existence of the objects which are, in fact, its members' (op. cit., p. 14). Hence Lombard really needs (MR) and (CE) to express his view about sets, while the supervenience principle (E) appears to be a consequence of the idea that the intrinsic features of an event which give its individual essence are its subject, time, and type, rather than a principle which leads to this view.

potential to have occurred earlier. Such a *de re* sceptic may say with some plausibility that although there are worlds which are distinguished in terms of where, when and how Socrates dies at them, there is no such distinction as the one Lombard aims to capture.

For the *de re* sceptic, Lombard's events are philosophers' fabrications; Lombard has done no more than isolate three features of events (we assume the type of an event will also be said to be essential to it) and attribute to the event the transworld identity conditions of the set of those features. This enables us to make sense of the *de re/de dicto* distinction as Lombard wishes us to, but only because his criteria have stipulatively *introduced* an entity with transworld being; only then can such a sentence as, say, 'the death of Socrates might not have been the death of Socrates', be given an interpretation; pre-theoretically, it is uninterpretable or obviously false. The source of such *de re* scepticism is not hard to seek out. It lies in the fact that ordinary event sentences of the sort that occur in non-philosophical discourse (this excludes e.g. 'every event is identical to itself') invariably permit paraphrase by other sentences in which the appearance of reference to an entity, the event of such-and-such happening, is eliminated. Thus, instead of saying that the assassination of Kennedy was the work of a conspiracy, we can say that there was a conspiracy to assassinate Kennedy; instead of saying that there were three attempts to scale Everest by the south-west face before the first successful one, we can say that three times it was unsuccessfully tried to scale Everest, where 'three times' is a temporal operator with a rather obvious evaluation clause in standard tense-logical semantics; and so on.

A full defence of such *de re* scepticism would require the production of a translation procedure, like the one which has been given here for translating a wide range of possible worlds sentences back into modal language; and, at present, there is little likelihood of such a scheme, since it is unclear what the procedures are which we employ to come up with paraphrases of particular event sentences, and unclear how rich the base language would have to be to permit the translation of all the event sentences we would want to be able to interpret. However, there is a well-known case to which the *de re* sceptic can point as illustrating what he has in mind here, the case of what Quine calls 'virtual set theory'.[20] Many statements which use terms of the form "the set of F's", i.e. statements which would naturally be formalized with set abstracts, can in fact be formulated equivalently without any apparent

[20] See Chapter 1 of Quine [1963].

reference to sets. For instance, the statement "the set of F's is a subset of the set of G's", i.e.

(a) $\{x : Fx\} \subseteq \{x : Gx\}$

makes inessential reference to sets, for we could simply say "all F's are G's",

(b) $(\forall z)(Fz \rightarrow Gz)$

instead. There is a principle at work here which is generalizable as follows. The language of sets contains a special predicate '$\in$' for set membership and a special variable-binding operator, the set abstraction operator (–:–). To permit the reverse translation of set-theoretic sentences, then, we need to be able to eliminate both the special predicate (which, since it makes sense only when applied to sets, would not be wanted in the base language) and occurrences of set abstracts. And, up to a point, this can be effected by the following rule:

(c) $r \in \{s : \phi\} = \phi[r/s]$

where r and s are variables, r and s not bound in ϕ, and $\phi[r/s]$ is ϕ with r substituted throughout for s. Recalling that (a) abbreviates

(d) $(\forall z)(z \in \{x : Fx\} \rightarrow z \in \{x : Gx\})$

it is easy to see that (c) yields (b) for (a). It is also easy to check the translatability of other simple expressions, and thus of more complex formulae built out of them; for instance, "the set of F's is the same as the set of G's" becomes "all and only F's are G's".

The limits of this translation procedure are reached when we consider set-theoretic statements in which set abstracts occur on the left hand side of '$\in$'. It is certainly possible to go on eliminating abstracts, by using

(e) $\{x : Fx\} \in \{x : Gx\} = (\exists z)(z = \{x : Fx\} \;\&\; z \in \{x : Gx\})$

and then applications of (c) and the biconditional treatment of set identity. But (e) introduces an existential quantifier, which says that something is identical to the set of F's; therefore (e) cannot be employed by one who seeks to show an ontology of sets to be eliminable.[21] However, so long as we restrict our theory of sets to those assertions

[21] Here I am assuming the standard interpretation of '$\exists$'. If '$\exists$' is given the 'semi-substitutional' interpretation, this would have to be qualified. See Parsons [1971].

for which an adequate reverse translation based on (c) is available, then we may reasonably be said not really to have an ontology of sets: thus Quine's use of the adjective 'virtual'. And provided our sets are virtual entities only, ones which we introduce by introducing certain means of expression subject to (c), we are not compelled by the arguments of earlier chapters to settle for any one particular account of their essences: there is no fact of the matter about the modal properties of these virtual entities. Of course, this is not to impugn the theory of essences which we have advanced for sets, since the conception of set we outlined, the iterative conception, for which the essentialist theory was developed, goes well beyond any conception consistent with sets being virtual.

In the case of events, it is conceivable that some of the technical terms of Lombard's theory are analogous to '∈' in preventing reformulation of sentences apparently involving an ontology of events. But this would have to be shown in some detail: given the variety of non-event-invoking means of expression illustrated three paragraphs back, there is initial plausibility in the thought that, even for such terms, clauses analogous to (c) will be available; indeed, Lombard's canonical form for terms referring to events might be a key component of the required reverse-translation scheme. If this is so, then the kind of distinctions between this very event's occurring in a world, on one hand, and an event just like it in certain respects occurring in a world, on the other, will not be a distinction he is entitled to, since all real distinctions between worlds will be manifested at the level of the paraphrasing sentences. From this point of view, what Lombard has done is just to erect a *de re/de dicto* distinction upon differences manifested at this level, and while this enables us to assign determinate truth-conditions to sentences regimented as *de re* modal sentences quantifying over events, there is no fact of the matter at dispute between Lombard and someone who chooses to erect the *de re/de dicto* distinction in a different way, a way which results in disagreement with Lombard's assignment of truth-conditions to those regimented sentences.[22] If what has

[22] Every non-*de re* sceptic would agree that Socrates' actual death occurs in a world w just like the actual world until after the time of his death, whereupon Socrates is resurrected to die again, since by the branching conception there is no real application of transworld identity in making this judgement. But if a *de re* sceptic about events appeals to anti-realism about them to justify his scepticism, and then gives a 'best candidate' criterion for their transworld heirlines, a criterion which would allow an event to change its subject, time, and type from world to world, it is hard to see that Lombard has presented any consideration

gone before is at all correct, there is thus the greatest possible difference between events, on the one hand, and material things and non-virtual sets, on the other.

which convicts such a sceptic of making a *mistake*. Evidently, the claim that there is the greatest possible difference between events and ordinary objects in this respect rests on the assumption that anti-realism about events is a position which is open to us, while anti-realism about common-or-garden objects is not.

9

The Justification of Modal Concepts

1. *Non-cognitivism*

IN THE foregoing chapters, we have investigated a variety of modal theses whose formulations all involve application of the concepts of broadly logical possibility and necessity, but we have at no point queried the legitimacy of these concepts themselves (as opposed to certain interpretations of them, e.g. the quantifier readings of '□' and '◇'). Since philosophers have in fact expressed scepticism about whether there is any well-defined content to be attributed to these concepts,[1] it would be unsatisfactory to conclude our discussion without addressing this issue, especially as it is not unreasonable to expect that our attempts to establish claims which employ these concepts should have increased our understanding of their nature, which in turn should help us expedite a defence of them. Nevertheless, it should be emphasized from the outset that this chapter is of a much more provisional nature than any of the preceding ones; specifically, at some points we will be able to do no more than describe problems for the approach we will take, since to tackle those problems adequately would require another book.

A justification of concepts of broadly logical modality should have two components, metaphysical and epistemological respectively. The first, metaphysical, component is a theory of what it is which determines the modal status of truths and falsehoods; a biologist can say what it is which makes 'Elizabeth II is descended from George VI' true, but it needs a philosopher to explain what feature of reality makes it *necessarily* true that Elizabeth II, if she exists, is descended from George VI. The second, epistemological, component comprises an account of how we come to know what is necessary and what is contingent, and by requiring this second component, we impose a constraint

[1] See Quine [1966, paper 13].

on the first: no metaphysical account which renders it impossible to give a plausible epistemological theory is to be countenanced.

A philosopher who holds that in some respectable sense there are features of reality which 'make modal judgements true or false' may be termed an *objectivist* about modality, provided it is understood that there is nothing in the use of this label which implies that an objectivist cannot appeal to facts about psychology; both 'internal' and 'external' features of reality are *prima facie* suitable for an objective grounding of modal truth. The position which affords the proper contrast with objectivism is that of the non-cognitivist, according to whom the content of modal propositions is such as to render the notions of truth and falsity not genuinely applicable to them. The best known non-cognitivist theory of a modal concept is Hume's theory of physical or causal necessity. According to Hume, if a subject undergoes repeated experience of C-type events being followed by E-type events then C-type events acquire the capacity to induce certain states of mind in the subject, and the feeling of this state of mind arising on perception of a given C-type event prompts the subject to hold that the particular E-type event which he perceives to follow, does so as a matter of causal necessity. On one construal, Hume's idea is not that the belief that a particular e follows a particular c of necessity can be *analysed* as the belief that e's following c is accompanied by the arising of the relevant state of mind, for that would be an objectivist view, since it would be a matter of fact whether or not that state of mind did arise; rather, his doctrine is that the content of the first belief which distinguishes it from the belief merely that e follows c is a component which *expresses* (without asserting) the fact that the subject is in the appropriate state of mind; in Dummett's terminology, when we say that c causally necessitates e, we are making a *quasi*-assertion rather than a genuine assertion.[2] Of course, in speaking of the repeated experience-types 'giving rise to' a capacity in future instances to 'induce' states of mind, double use is being made of the concept being explained; but there is no circularity here, since the explanation can be applied over again to the terms of the theory itself.

Hume did not much address himself to the question of either strictly or broadly logical necessity, but one can see, at least very roughly,

[2] See Dummett's distinction between assertions and quasi-assertions, [1973, pp. 352–355], where he writes: '. . . we may well suspect that such [non-cognitivist] theories represent rather cheap attempts to resolve difficult philosophical problems by ruling them out of order.'

how he might have extended his account of causal necessity to stronger modalities. When we assert 'Necessarily, A' we would be said to be asserting A and expressing the fact that apprehension of the proposition expressed by A induces a certain state of mind in us. The capacity of the proposition to induce the state of mind would be one it has in virtue of some feature of it with the role which corresponds, in the case of causal necessity, to that of the regularity in the subject's experience of C-type events being followed by E-type events; this feature is the one which causes the impressions in us from which we derive the concept of necessity. But what feature of propositions endows them with this power? Hume comes close to addressing this question in the following passage, noted by Stroud:

> Thus as the necessity, which makes two times two equal to four, or three angles of a triangle equal to two right ones, lies only in the act of the understanding, by which we consider or compare these ideas; in like manner, the necessity or power, which unites causes and effects, lies in the determination of the mind to pass from the one to the other.[3]

Hume's view is that we then make a mistake: we project something essentially 'inner' onto the external world, and come to the mistaken belief that the concept of necessity we have applies to propositions in virtue of objective properties of ideas and, as a consequence of this, we mistakenly believe that modal judgements can be true or false.

We can explain this further by reference to the primary/secondary quality distinction. We have already noted that there are two distinct ways of drawing this distinction, either in terms of whether or not a property is a disposition to produce in us sensations with certain features, or in terms of whether or not the property is one which is really 'in' the object itself, as opposed to being possessed by the object only in virtue of features of sensations resulting from perception of the object. Up to a point, Hume's view of necessity is like the view of secondary qualities held by someone who draws the primary/secondary distinction in the second way, but the comparison is slightly misleading, for such a person need not be a non-cognitivist about secondary quality terms; he can say that it is literally true that an American mail-box is blue since it is literally true that perception of such a thing has certain effects on us, while Hume, on the present construal, denies that it is ever literally true that a proposition is necessary, since this does not *mean* that the act of the understanding consisting in our reflecting on

[3] See Stroud [1977, p. 241].

it has certain effects on us. The relationship of the effects to the content is rather like a Humean view of the relationship of the effects of perception of a distasteful scene to the content of a judgement of moral disapproval about the relevant goings-on: the effects are certain psychological states in us, and part of the content of the judgement, the 'evaluative' part, expresses but does not assert the fact that we are in such states, being in which motivates us to behave in certain ways by producing desires or aversions in us, the exact nature of which turns on contingent features of our psychological make-up. So a non-cognitivist might try to elucidate the difference in content between 'A' and 'Necessarily, A' in terms of the attitudes towards the proposition A which arise in us upon the obtaining of the psychological states whose presence is expressed or signalled (in the way that a cheer expresses or signals approval) by the use of the modal prefix.[4]

Another well known non-cognitivist account of a modal concept is Wittgenstein's theory of mathematical necessity (here I follow the interpretation of Wright[5]). Wittgenstein was an extreme conventionalist about mathematical necessity, in that he held that there is no sense in which the premisses of a mathematical proof necessitate its conclusion: any claim of necessity merely lays down a new convention. His position contrasts with that of the moderate conventionalist, who holds that once certain general conventions governing the logical constants are accepted (rules of inference) together with certain other conventions about non-logical words (basic axioms) the conclusion of any proof is necessitated by its premisses in virtue of each step's being an application of some already accepted general convention; thus any necessity which is not immediately a matter of convention ultimately reduces to conventions. The difficulty for the moderate conventionalist is to explain in what sense the correctness of a particular step in a proof is a 'consequence' of an antecedently accepted convention: if this is a further convention, we have an apparently unending regress of conventions, while if it is not, we have a relationship of necessary consequence which is not subsumed under any convention. The difficulty for Wittgenstein's view, on the other hand, is to square it with obvious facts about mathematical practice, especially the feeling of any competent person following a proof that he is *compelled* to accept each new line as a consequence of the earlier ones.

[4] See Stroud [1977, p. 82] and also Peacocke [1982] for discussion of contemporary versions of Hume's non-cognitivism about causal necessity.

[5] See Wright [1980], especially chapters 19–23.

The view which Wright extracts from Wittgenstein is that each mathematical theorem is like a rule of a game: the rules themselves cannot be said to be true or false, and as rules are added (in the hope of producing a better game), new constraints are thereby imposed on what is admitted as correct play. But we add new rules to a game as a matter of *decision*, so by analogy Wittgenstein held that there is an element of decision in accepting a new proof in mathematics. Explicit assertions of necessity he interprets as prescriptive: to say that it is necessary that p, is to urge a certain policy, namely, that of not admitting any description such that were some situation to be accurately described in that way, its obtaining would refute some non-modal generalization 'appropriately' related to p. For example, to say that it is necessary that $7 + 7 = 14$ is to urge, amongst other things, that we do not ever accept 'there are 7 apples in A's basket and 7 apples in B's and 15 in both', since this conflicts with: if there are 7 F's in one container and 7 G's in another, then there are 14 F-or-G's in both. Since prescriptions of policy are not true or false, neither is any sentence of the form '□P'.[6]

A non-cognitivist theory of apparently fact-stating discourse is usually a 'last resort' account, deriving from some general doctrine about meaning. To the degree that one finds the non-cognitivism unbelievable, one will prefer to reject the general doctrine but, of course, the *right* to find the non-cognitivism unbelievable must be earned by a critique of the semantic theory (or in Wittgenstein's case, semantic scepticism) underpinning it. Perhaps no one is inclined to Hume's semantics nowadays; but addressing Wittgenstein's scepticism is another matter, and is one of the tasks we mentioned earlier as requiring a separate book. At this point, therefore, we will just continue the taxonomy of accounts of modal concepts at variance with the one we will eventually adopt.

2. *Quasi-psychologism*

Since the view we will investigate next does not claim to provide an *analysis* of possibility and necessity, we call it quasi-psychologistic rather than psychologistic; nevertheless, the main idea of the approach is that the modal status of truths and falsehoods is ultimately grounded upon human intellectual abilities. For example, Stroud has written that to explain necessity and possibility we need to pursue 'questions about

[6] Wright provides a useful summary of his interpretation in [1980, pp. 410–11].

the human mind and its capacities'; and these explanations must appeal to 'empirically discoverable, natural facts about us'.[7] One interpretation of these claims, the preferred one, is that they are of a piece with the Wittgensteinian non-cognitivism sketched above. Wittgenstein suggests that it is a matter of decision whether or not to accept the conclusion of a proof, yet it is clear that there is no disagreement between trained subjects on whether or not a given proof should be accepted (modulo the constructivist/platonist dispute), which may seem surprising if *decisions* are called for. But the phenomenon of agreement can be explained by the brute propensity of humans who have been trained in certain ways all to go on in the same way, a propensity itself residing in, or being an, empirically discoverable natural fact about us. However, it is also possible to interpret Stroud's remarks as advancing the view that the necessity of a truth follows semantically from our inability to make sense of the supposition that it is false, an inability which itself admits of an explanation which will not cite features of the subject-matter of the proposition in question, so that facts about the mind's capacities are genuinely the basic ones. Thus modal statements will be true or false depending on facts about our abilities.

Wiggins has explicitly urged a quasi-psychologistic approach:

> What we have here . . . is not a reduction or elimination of necessity or possibility . . . but the following elucidation of possibility or necessity *de re*:
> (i) x can be ϕ iff it is possible to conceive of x that it is ϕ;
> (ii) x must be ϕ iff it is not possible to conceive of x that it is not ϕ.[8]

So, according to Wiggins and the second reading of Stroud, the first step in an analysis of possibility and necessity is to look to what we can and cannot make sense of, or conceive (in some sense of 'can' and 'cannot' other than the broadly logical, one assumes).

It seems that this approach is either circular or else extensionally incorrect: there are some impossibilities which it classifies wrongly. To see this, let us agree with Wiggins that no android could have been human, and conversely. Suppose now that we are visually presented with a creature which for *all we know* may be either android or human, but which is actually an android. Then it is epistemically but not logically possible that this creature is human, and so one might say that conceivably it is human; thus, *ad hominem*, Wiggins is wrong to equate what is conceivable and what is possible. However, this argument turns

[7] Stroud [1977, p. 245]. [8] Wiggins [1980, p. 106].

on interpreting 'conceivable' as 'epistemically possible', which may be at once too broad and too narrow an interpretation. It is therefore reasonable to ask that this sense of 'conceivably' be put aside as not what Wiggins means (nor what Stroud means, if we identify what we can conceive of with what we can make intelligible to ourselves).

There is another sense of 'conceivably' which is not so straightforwardly tied to epistemic considerations, and which, again, is not what Wiggins should mean, since the impossible is also conceivable in this sense. Kripke sometimes writes that we can conceive of finding out that such-and-such is the case, even though we in fact know that it is not the case, and can infer from that that it is *necessary* that it is not. Moreover, although 'conceivably' is intensional, it does seem that if we can conceive of finding out that p then we can conceive of p's being the case: someone who holds that p is inconceivable, if he also agrees that p's being found out implies its being true, could at best conceive of a person claiming to have found out that p, or of an investigation which results in such a claim, a claim which then survives our best efforts to refute it; we can say this without endorsing the false general principle that if I is an intensional, and F a factive, operator, then IF(p) implies I(p) (consider 'Smith hopes that Jones knows that p'), the point being specific to the meanings of 'it is conceivable that' and 'it is discovered that'. Then, since we can certainly conceive of finding out that the android is human, or better (in Wiggins's style, since 'the android is human' is a *de dicto* impossibility), since we can conceive of finding out, concerning the android, that it is human, then we can conceive, of the android, that it is human, even though we know it is an android.[9]

An explanation of how this is possible should be sought. The explanation is that, in this sense, 'It is conceivable that p' means that the hypothesis that p does not by itself contradict any principle which is constitutive of the content of a concept involved in the proposition that p: refusal to rule out p *a priori* is *not* indicative of failure to grasp some of these concepts. (In fact, there are really two explanations here, depending on whether we require that there should in fact be no such contradiction, or just that none should be evident to the subject, but in this context the ambivalence is irrelevant.) Thus, in the example, we can conceive of the android that it is human because the android is presented to us perceptually, and it is not required to

[9] Examples of Kripke's remarks about finding out are [1972, pp. 269, 313–14, 318–19].

have a demonstrative thought about it that we think of it as an android, even if we in fact know that it is one. Someone who holds that it is necessary that water is H_2O can allow that it is conceivable that water is not H_2O for a similar reason, for a man can have a way of thinking of the substance water which permits him to have thoughts about water and which does not at all involve him thinking of water as H_2O; in view of the way use of the word 'water' is taught, by ostension of samples, most persons' concept of water will be like this. And, in the most familiar example, even after we have discovered that Hesperus and Phosphorous are the same planet, it is still conceivable that they are distinct, because we acquired two different ways of thinking of that planet which, even when they are conjoined, do not imply the identity.[10] If we say that p is *strongly a priori* iff failure to assent to p is indicative of failure to grasp some concept involved in the thought that p or failure to perform some elementary logical inference, then the present senses of 'conceivably, p' are comprised in 'it is not strongly *a priori* that not-p'.

There is one special case where we might expect a close relationship between conceivability and broadly logical possibility, the case of sets. Wiggins's acceptance of essentialism about set-membership but scepticism about the necessity of origin arises out of his attempt to associate conceivability and possibility, for it is plausible that when we think of the set $X = \{a, b\}$, we do indeed think of it as the entity whose members are exactly a and b, and it is therefore inconceivable that it should lack these members or have any others. But in this sense of 'conceivable', as we have just seen, the other component of Wiggins's essentialism, according to which it is of the essence of a thing to be of the kind of which it actually is, fails to follow: we can conceive, of the android, that it is human. And it is plausible that attempts further to refine conceivability to get rid of this consequence will bring with it other essentialist claims about which Wiggins is sceptical, such as the necessity of origin.

The main problem for Wiggins's approach is to come up with some further sense of 'conceivably' which does capture broadly logical possibility, a sense which can be characterized, if not in the fundamental

[10] To say that a way of thinking of a planet is associated with a particular name of it is not to suscribe to a description theory of names, if that is a theory about what it is for an object to be the referent of a name. What makes an entity x the intentional object of a way of thinking will have to do with the relations in which the thinker stood to x in acquiring that way of thinking of it. See Evans, [1982, pp. 14–22].

terms of the theory of psychological capacities, at least in terms which are sufficiently far removed from those which a non-psychologistic objectivist might himself use to explain broadly logical modalities. But in the prevailing absence of any detailed quasi-psychologistic theory, it is very hard to see where such a characterization of 'conceivably' is to come from. The etymology of the word demands that at least consistency of concepts be imposed, but we have seen that a further restriction is required, and there is apparently no other psychological capacity which will exclude just what needs to be excluded; for instance, the faculty of pictorial imagination is powerless to exclude the conceivable impossibilities: we can certainly *picture* discovering, of the creature in front of us, that it is human. Thus the whole approach in terms of conceivability looks unpromising.

3. *The theory of content*

The explanation of the second sense of 'conceivably' in the previous section would be regarded as useless by many philosophers, on account of its appeal to principles which are supposed to be constitutive of the content of the concepts appearing in the hypotheses asserted to be conceivable. The problem is with the notion of the 'content' of a concept, a notion said by Quine, for instance, to be itself without content. Since we are going on to attempt to explain necessity in terms of the content of concepts, we must at this point address, or at least note, Quine's views. An *a priori* principle constitutive of the content of a concept, say the concept of being an F, is intended to contrast with *a posteriori* beliefs about F's, which are beliefs whose possession requires prior mastery of the concept. However, Quine has argued influentially that this alleged distinction between *a priori* truths about the concept of F-hood and empirical truths about F's is not one which survives careful scrutiny.

Anyone with a broadly empiricist outlook who tries to maintain the distinction will attempt to do so by taking the *a priori* principles to be principles about how application of the concept should be constrained by experience (perhaps relative to the application of other concepts), but, according to Quine, although this is what such propositions *should* be like, there are no propositions which express conceptual truths in this sense. For, if there were, then given a sequence of experiences recalcitrant with respect to our current views about F's, those conceptual truths, if they really are such, should dictate which of the empirical propositions about F's that we currently believe should be abandoned, and which empirical propositions about F's we should come to believe instead; or, at least, they should dictate that certain revisions are open while certain others are not, even if they do not determine a unique candidate. But Quine argues that, in fact, no particular propositions about F's have this role: the impact of any sequence of experiences can be distributed throughout the range of propositions a thinker may be disposed to assert prior to undergoing that sequence, in an endless variety of ways. For example, a scientific theory may consistently be held true regardless of the evidence to the contrary, if the theorist is willing to continue to append *ad hoc* hypotheses and complicate other parts of his theory to explain away the awkward evidence: this procedure could be carried even to the point of abandoning logical principles. Thus, given some proposition purportedly stating an *a priori* principle constitutive of the concept of being an F, we could choose a revision of our theory of F's to accommodate some experiences of F's, which revision involves a rejection of the alleged constitutive principle.[11]

In assessing the plausibility of Quine's views, a task which cannot be pursued very far here, one should separate the relatively uncontroversial idea that experience confronts whole theories rather than single hypotheses on a one-to-one basis, from the much more controversial view that any adjustment of a theory is open to us in advance. Someone who accepts the holism of the first thought is no more committed to denying determinate content to the individual hypotheses which comprise the theory than is someone who holds that individual hypotheses are the primary bearers of content committed to denying determinate content to the words which make up the hypothesis. In each case, the content of the part can be identified with its contribution to the content of wholes in which it may occur. The second element of Quine's view decrees that nothing can be isolated for individual hypotheses as comprising such a contribution, but this appears to be problematic. First, it is unclear that sense can be given to the notion of an experience's being *recalcitrant* for a given theory if no adjustment is ruled out in advance; the logical principles which determine what is recalcitrant and what is not seem very different from the working

[11] The classic statement of Quine's position (qualified in later work) is of course 'Two Dogmas of Empiricism', on which this account of his views is based. See paper (ii) in Quine [1961]. In the comments that follow, I am indebted to part I of Dummett's 'The Significance of Quine's Indeterminacy Thesis'; see Dummett [1978, pp. 375–84].

hypotheses of the theory itself. If one can dissolve recalcitrance by treating these principles as if they were mere hypotheses and abandoning them, one begins to lose one's grip on what the goal of inquiry is and why it is pursued at all. Secondly, Quine's no-specific-bearing doctrine seems to contradict the facts about the practice of science, for in most cases scientists do not have much difficulty in determining to which part of a given theory particular experimental evidence is most relevant. But perhaps philosophers of science have only recently begun to develop the sophisticated analyses of the relation of evidence confirming or disconfirming a theory which would be required for a full answer to Quine on this point.[12]

Someone who proposes that the content of a hypothesis can be identified with its contribution towards the content of any theory to which it belongs is advancing at best a schema of a position, until he provides some identity criterion for contents to articulate the kind of contribution which he has in mind. An interesting proposal to this effect has been made by Hartry Field.[13] Note that for Quine's point about different ways of revising a theory to obtain, we have to be considering two theorists, or a single theorist at different times; for, obviously, one theorist can in one revision revise a theory in only one way. So we might suggest that for a subject S at time t, propositions p and q are the same iff the experiences of S through t either warrant (for S) the holding of both or fail to warrant (for S) the holding of both. But this is clearly far too simple, since it implies that all the propositions S holds at t have the same content: we also wish to consider what S would say about p and q under the supposition that his experience is like *this*, or like *this*, etc. Moreover, the division of propositions into two classes by a given sequence of experiences, those warranted by it and those not, is too coarse. Experience confirms or disconfirms hypotheses, and confirmation comes in degrees. Field combines these points into the following criterion of sameness of content (in his terminology, sameness of conceptual role) of p and q for S at t: p and q are the same iff for any proposition r, S's subjective conditional probability for p given r is the same as his subjective conditional probability for q given r. That is,

(1) $p = q \quad \text{iff} \quad (\forall r)(Pb(p|r) = Pb(q|r))$[14]

[12] Here I have in mind especially Glymour [1980]; see pp. 110–23.
[13] See Field [1977].
[14] See the definition of equipollence at p. 382 of Field [1977].

where Pb is S's subjective conditional probability function at t; Pb(x, y) is the probability x has for S given y (which need not be the probability S would ascribe to x were he to come to believe y).[15] We might have tried to restrict the range of the variable 'r' to 'observation propositions', that is, to propositions which would merely state how things look or have looked to S, but (1) allows for the conceivable case in which S's subjective conditional probability function is such that *distinct* non-observational propositions are assigned the same probability come what experiences may, but whose conditional probabilities come apart given some non-observational proposition.

Criterion (1) leads to an elucidation of the relativized notion of a principle constitutive of a concept for a subject S at a time t; this would be a principle which, for S at t, has maximum probability regardless of what proposition r is given, and which is relevant to the account of S's reasons for making the assignment of conditional probabilities he does in the cases of propositions which involve the relevant concept and which, for some r, are not maximally probable given r. Here we have no very radical departure from Quine; for instance, as time passes, it is still open to us to say either that the subject is changing his beliefs, or that he is altering the content of his concepts.

Criterion (1) may appear obviously circular, since in attempting to explain identity for propositions, it quantifies over propositions, so that if the right-hand-side of (1) were applied in an attempt to settle an identity question, the verdict could turn on whether or not univocal substitution for r is being made, which in turn could depend upon whether or not p and q *are* the same proposition. However, a similar situation arises with a number of 'synthetic' identity criteria for categories of entity: that material objects x and y (of the same sort) are identical iff for any material object z and time t, x is in spatial relation R to z at t iff y is in spatial relation R to z at t; that events e and f are the same iff for any event d, e is a cause (effect) of d iff f is a cause (effect) of d; the functionalist criterion for identity of mental states is also similar in structure. Such criteria merely specify a relational framework within which we individuate the relevant entities in a manner consistent with the criterion; the criteria give the terms in which we specify that in which identity and difference for those entities consists.

[15] The s.c.p. of 'this die will show a three' given that it will show either a three or a five is, for most people, ½. But the actual world might be such that if I were to come to believe that the die had shown a three or a five, then I would acquire additional beliefs as a result of which I would not agree that the chances of it being three are 50–50.

To give a non-relative account of the distinction between the conceptual and the empirical, therefore, what we have to do is to specify a relational framework which comprises the thoughts of different subjects, or of the same subject at different times. We can isolate some problems which face us here by considering why Field's criterion cannot just be generalized in the most straightforward way: the obvious objection is that two subjects may well attach the same content to some proposition but disagree about conditional probabilities for it since they disagree about background facts. But there is a natural way to avoid this objection, for if the two subjects do attach the same content to some proposition, we would expect this to be manifested counterfactually: if they were to agree in their background beliefs, they would agree in their assignments of conditional probabilities to that proposition. So the suggestion is that for any propositions p and q such that S believes p and S′ believes q, we should say:

> p = q iff for any collection of background beliefs B, if S and S′ were both to accept B, then for any r, it would be that pb(p|r) for S = pb(q|r) for S′.

As it stands, this criterion is in need of explanation and refinement. For example, for each proposition p, we have to find some way of circumscribing what facts are background relative to p, so as not to include in the 'background' e.g. S's beliefs about the conditional probability of p, for each r. But the more pressing question is whether there is reason to hold that, in principle, no such counterfactual criterion can succeed. The problem is again one of apparent circularity, since we are presuming on the notion of S and S′ having the *same* background beliefs B, and also using the variable r to stand for one and the same proposition as the given condition for the assignments both S and S′ make. The presumption of same background beliefs certainly imports an extra element of complexity to this criterion over and above what was present in the earlier, intrasubjective criterion of identity at a time, and raises questions we cannot possibly pursue adequately in the present context.[16]

There is a further problem with the present line of inquiry, having to do with the suitability of the notions being employed for substantiating a notion of the content of a concept our grasp of which

[16] In the last paragraph, I have been much indebted to Peacocke's unpublished [1983b], where the crude counterfactual identity condition given above is refined to meet many of the objections to it.

permits us to arrive at principles constitutive of that concept. As Wright has written:

We want to attribute to ourselves a capacity reflectively to apprehend impositions and constraints which the manner in which we understand particular expressions places upon us . . . the capacity is thus . . . essentially a capacity to discern the character of one's own understanding. [But] Wittgenstein repudiates the view that each of us may regard himself as knowing reflectively what kind of application of an expression conforms to the meaning he attaches to it.[17]

As with his conventionalism about mathematical necessity, there is again the problem of squaring this claim with the subjective phenomena. Thus a philosopher, in investigating whether a causal judgement is always equivalent to some related counterfactual, may consider a case where the relevant counterfactual is true, e.g. 'if his sister had not had a child, Smith would not have become an uncle', and conclude that there is no equivalence, since the application of 'causes' which conforms to the meaning he attaches to it is that his sister's giving birth did not cause Smith to become an uncle. Wittgenstein would claim that we just find ourselves with a brute propensity to say one thing rather than the other, and that the hypothesis of a capacity to apprehend the content of one's concept of causation and employ that apprehension in the testing of philosophical analyses, does no work. But we do have the practice of testing an intuition against a range of cases in a search for consistency or for an answer to a given hard case, which it is natural to describe as trying more accurately to apprehend the content of the concept in question. And we are familiar enough with how children acquire conceptual sophistication, e.g., how a sequence of question-and-answer sessions in the presence of observable phenomena can lead a child to realize that 'x arrives at the destination before y' is insufficient for the truth of 'x travelled faster than y'. With these phenomena in mind, the entities whose identity condition are given by Field-style criteria seem appropriate objects of reflective apprehension, since what we apprehend is a difference in our reactions to a real or imagined situation, according to which proposition we are entertaining as assertible.

Is our drawing one conclusion rather than another anything more than the manifestation of a brute propensity? Appeal to brute propensities can always be made to explain any behaviour whatsoever, but if we are not to be denied at the outset the right to ascribe some mental

[17] See Wright [1980, pp. 354–5].

life, e.g. beliefs, desires, and intentions, it is unclear why we cannot ascribe states of understanding to explain the kind of behaviour just described. Perhaps the conventionalist theory of necessity can be made to account for such phenomena as the search for consistency, but that theory was supposed to be ushered in only *after* the critique of such notions as 'the content of a concept' had done its work and left us looking for a new way to draw the necessary/contingent distinction. Wittgenstein's critique is to the effect that for a word to have a definite content is for there to be a distinction between correct and incorrect application of it, while if meanings are cognitively accessible in a special way from the first person point of view, then we are not even in a position to draw a distinction between a word's having a determinate meaning and its having no meaning at all, so that applications of it are quite arbitrary. A tremendous weight is therefore borne by this contention about an individualistic conception of states of understanding (as sketched in the quotation from Wright). Certainly those philosophers who have seen no paradox in allowing that a subject can believe that he has a particular singular belief, although in fact he has no such belief, have not felt compelled to abandon individualistic conceptions of what it is to hold a belief.[18]

Indisputably, much more would have to be said at this point to provide a genuine vindication of the traditional notions.[19] But perhaps enough has been said to establish the following modest rationale for moving on: however difficult the issues raised by Quine and Wittgenstein are, the assaults on the traditional notions are not so immediately compelling that all interest in a justification of modal concepts which employs these notions instantly evaporates. So we will proceed with the development of such a justification.

4. *The source of necessity*

The striking feature of the arguments we gave in earlier chapters in defence of such *de re* modal principles as Crossworld Extensionality and the Necessity of Origin is that they are wholly *a priori*: the doctrine which does most of the work is that identity is an intrinsically grounded relation, and this doctrine, if true, is true in virtue of the content of the concept of identity, and is established by *a priori* reflection upon that concept. This suggests that we can explain the

[18] See Evans [1982, pp. 44–6].

[19] I have discussed one interpretation of Wittgenstein's critique of these notions in Forbes [1984a].

necessity posited in the principles as arising out of *a priori* facts about the content of the concepts involved in them. However, to make this more precise, we must explain carefully how the necessity arises, for its having its source in the content of concepts has to be shown to be consistent with certain other phenomena; in particular, with the conceivability of the opposite of something metaphysically necessary (for we already explained such a sense of conceivability in terms of consistency with conceptual content); and with the obtaining of necessary *a posteriori* and contingent *a priori* truths.

This last phenomenon is in fact not one which presents much of a difficulty, provided one agrees that the 'canonical' or most direct method of establishing a necessary *a posteriori* truth is by inference from a singular *a posteriori* truth and a general *a priori* one; for then the source of the *necessity* in an *a posteriori* truth is still an *a priori* truth. Certainly, all the familiar examples are like this; for instance, the necessity of Hesperus's being Phosphorous is inferred from the hypothesis that Hesperus *is* Phosphorous, itself based on inference from physical theory and observational evidence, together with the necessity of identity, which is defensible only *a priori*, if at all. We may conjecture that no necessary *a posteriori* truth departs from this pattern, and might expect an account of the source of broadly logical necessity to have such a consequence. So far as the contingent *a priori* is concerned, it has been well argued elsewhere that the contingency of such statements is in a good sense superficial, and our account of the *a priori* grounds of necessity will be consistent with superficial contingency in *a priori* truths, since that is consistent with the *a priori* truths still giving rise to non-contingency of a 'deeper' sort.[20]

Broadly logical necessity may be *de dicto* as well as *de re*, but the gulf between *de dicto* necessity and conceivability in our second sense is apparently of a different nature from the gulf between the latter and *de re* necessity. *De dicto* necessities are straightforwardly explicable in terms of the content of concepts, for they are simply definitions, or principles constitutive of some concept's content, or logical consequences of some concept's content, or logical consequences of such principles. This is not to say that it cannot be a matter of controversy whether an alleged *de dicto* necessity really is such, for it may be a matter of controversy whether a given principle really is constitutive of a concept's content. This can happen even with logical concepts, where broader considerations about the nature of content, such as

[20] See Evans [1979] and Davies and Humberstone [1980].

those urged by Dummett in his defence of intuitionistic logic, need to be appealed to to judge the putatively constitutive nature of particular principles, in this case, natural deduction rules.[21]

Other disputes concern whether or not an alleged *de dicto* necessity can really be shown to be a logical consequence of content-constitutive principles. For instance, establishing that nothing can be both red and green all over, or even just red all over and green in part, requires an unobvious derivation of the mutual exclusiveness of colour classifications for a fixed surface area: such exclusiveness is not apparent from principles constitutive of the content of individual colour concepts. As is familiar, difficult questions about *de dicto* necessity are especially common in connection with the concepts of space and time. It may seem from these remarks, in fact, that there is no gulf at all between conceivability and *de dicto* possibility, for if conceivability requires logical consistency with constitutive principles, then it must be co-extensive with *de dicto* possibility. But if we recall our distinction within our second sense of conceivability, according to whether we require *logical* or merely *epistemic* consistency with constitutive principles, we can see that there is room for a gap between *de dicto* possibility and one of the distinguished senses of conceivability. For if only epistemic consistency is required, a *de dicto* impossibility may be conceivable, when the conflict of the impossible hypothesis with constitutive principles is not perceived.

The position which we will now argue for is that *de re* necessity does not differ from *de dicto* necessity in respect of how it arises: it is still a form of conceptual necessity. However, while a *de dicto* thesis wears its conceptual content on its sleeve, the concepts which are the source of the *de re* necessity are not manifest in the simple form '□Fa'. We can bring this out by contrasting the conceivability of 'not-Fa' with its impossibility. 'not-Fa' is conceivable because the only concepts principles governing it which must be respected (epistemically or logically) are those expressed by the predicate or involved in the way of thinking of the subject associated with the subject term; if the subject term is a perceptual demonstrative, these would be the concepts needed for a specification of the representational content of the perception, what it is 'as of'. But the truth of '□Fa', if it is true *a posteriori*, is to be explained by the involvement of further concepts.[22]

[21] See Dummett [1975b].

[22] The truth of *a priori de re* necessities about individual objects, for instance, '□—(Fa & —Fa)', can be explained simply by their being implied by *de dicto* necessities, or, as in this case, by being instances of a *de dicto* scheme.

It would be unilluminating to say that when '□Fa' is true, this is because principles governing certain concepts require that 'Fa' be true in every world, not merely because if 'Fa' is *a posteriori* this claim would be false, but because 'true in every world' simply repeats '□'; rather, we want the correctness of the attributions of necessity to be a consequence of the fact that certain conceptual relationships obtain. Furthermore, at this point we want an exposition free of the apparatus of possible worlds, since it is because the content of the modal concepts are as they are that this apparatus can be applied; so, until we have independently specified the content of the concepts interepreted by the extensional machinery, we shall not speak of the intrinsic grounding of identity or any other transworld relationship.

Let us recall what we said would be the standard form of an essentialist thesis,

$$(S') \ \Box(\forall v)\Box(\forall u_1) \ldots \Box(\forall u_n)\Box[(C(v) \,\&\, A(v, u_1 \ldots u_n)) \rightarrow \Box(E(v) \rightarrow A(v, u_1 \ldots u_n))].$$

True instances of this are *a priori* truths. Additionally, the category concept (expressed by the predicate substituted for) C and the concepts in the expression (substituted for) $A(v, u_1 \ldots u_n)$ seem, in the true instances, to be related as follows: our understanding of what it is to be a thing of category C involves, at least in part, thinking of it as a thing with certain properties, or standing in certain relations to some other things, where the relevant properties and relations are specified in the (instance of the) formula $A(v, u_1 \ldots u_n)$. For instance, one's understanding of what it is to be a set involves thinking of a set as a gathering together of antecedently given entities (this is the iterative conception of set). One's understanding of what it is to be an artefact involves thinking of an artefact as a functionally unified assemblage of components (or a single component) the form of which is fixed by some design. And one's understanding of what it is to be an organism, a living thing, involves thinking of an organism as an item with a characteristic biological functioning occupying some niche in a generational tree of *self-reproducing* entities.[23] So here we

[23] In [1979, p. 96] Maynard Smith offers this definition: 'We shall regard as alive any population of entities which has the properties of multiplication, heredity and variation'. Presumably the population has the properties in virtue of properties of the entities themselves, and I am taking it that the capacity to reproduce itself is a main characteristic of a living thing, together with its having been the product of reproduction, and thus having undergone growth from a starting point.

have *a priori* truths constitutive of the concepts of set, artefact, and organism.

It is surely no coincidence that those concepts which appear in the description of what it is to be a thing of a certain category are the characteristic concepts of the *a priori* essentialist theses which are the modal premisses of the canonical derivations of *a posteriori de re* necessities concerning things of those categories. However, what we have said so far is still sufficient only to explain *de dicto* necessities. To explain the necessities an essentialist posits, we must link the introduction of *de re* modality to the rigidity the essentialist claims for the properties and relations, concepts of which figure in the content of our understanding of what it is to be of that category. The heart of our proposal about the link is this. Mastery of the *de re* use of modal operators requires more than a disposition to employ them consistently with the interdefinabilities and the *de dicto* truths: there are many configurations of properties and relations in which a given sequence of individuals cannot stand, despite the satisfiability of the configuration by *some* sequence of individuals. The intuitive thought which we have about such cases is that any individuals which stood in that configuration would not be *these* individuals, a thought which implies a conception of what it is to be *this* individual, or *this* one, etc., a conception of something which does not alter under any counterfactual hypothesis which is itself genuinely possible; we can call this conception the conception of an individual's thisness (not to beg any questions against the Haecceitist here, we can allow that the conception may admit of articulation only by necessary conditions which are not jointly sufficient). The explanation of at least the *de re* necessities which instantiate (S′) suggested by these points, then, is that we form conceptions of thisnesses by invoking the concepts involved in how we think of what it is to be a member of a particular category, which concepts are either monadic or consist in certain existentially quantified conditions: having *some* members, *some* components, *some* starting point in a self-reproductive act of a parent or parents; our method is to fix the content of the thisness of an individual x in terms of the identities of the entities which, for the individual x in question, satisfy the conditions which are existentially quantified in the specification of our understanding of what it is to be a thing of x's particular category. That is, for any x, our idea of what it is to be x is that being x is being the thing which has the individual nature specified by the properties, relations and relata introduced in the manner just described by the category concept for the category to which x in fact belongs.

The natural response to an unreasonable hypothesis of possibility for an object x, that in such a state of affairs it would not be x which satisfied the conditions, is evidence that we do possess concepts of thisnesses for individuals. The necessity of some *a posteriori* truth about an object x may then be explained by its asserting that certain objects stand in certain relations, just the objects and relations which are specific to the individual nature of x. This is the simplest case, while more complex *a posteriori* truths will be necessary if they are modal logical consequences of simple *a posteriori* necessities and *de dicto* necessities. Our knowledge of how the specific content of the thisness of a particular individual follows from *a posteriori* facts about it is itself expressed in *a priori de re* principles such as the true instances of (S′).

However, this whole account is clearly tailored to the needs of the theory of essences which has been defended in previous chapters, and may therefore appear suspicously *ad hoc*. To remove this appearance, we need a reason why a conception of individual thisness is required and why it should be derived in a fashion similar to the one we have spelt out.

That *some* conception of individual thisness is required is no mystery, for if the practice of making *de re* attributions of possibility to objects is coherent at all, there must be a distinction between correct and incorrect attributions. So the question is why this distinction should be drawn in terms of the sort sketched above: why not settle for a boundary marked by conceivability in some sense or other, or by the property of not implying any *de dicto* impossibility? To see our way towards an answer to this question, it is helpful again to revert to the analogy with time. To make the step from *de dicto* uses of tenses to *de re*, we have to master a conception appropriate for the temporal case of what it is to be a particular object, a conception under which objects have determinate pasts and futures, just as the conception for the modal case is one under which they have determinate possibilities (perhaps to varying degrees). The temporal conception we employ is (at any rate in part) that of a thing as an occupier of a continuous route through space, a route which continuously unfolds with the passage of time. But why this conception?

Perhaps we can say a little more here than: this is what we do. Shoemaker has introduced the idea of a 'gerrymandered' object to canvass some alternatives; for instance, a 'klable' may be defined to be the object consisting in the stages of a certain table from noon to midnight and the stages of a certain chair from midnight to noon, and exists so long as both table and chair exist.[24] Then there could be truths about klables, since these would just be elliptical for truths about tables and chairs, but klables would not be on an ontological par with tables and chairs, according to Shoemaker, since the former are logical fictions while the latter are not; for Shoemaker, 'klable' does not pick out entities which are really there, awaiting linguistic recognition. Similarly, we would like to be able to hold that a use of *de re* modality in which the objectual quantifiers range over entities whose possibilities are circumscribed only by *de dicto* necessities does not identify ranges of possibilities for real things, in the way that the past and future of a klable is not the past and future of a real thing; in extensional terms, the transworld heirline of an 'entity' for which all is possible but a *de dicto* impossibility is no more the heirline of a real thing than the spatio-temporal path of a klable is the path of a real thing. And the same should be said about other less liberal attempts to circumscribe *de re* possibilities which conflict with the essentialist theses which have been defended in this book.

Suppose, however, that someone presses the question why there are no klables, and is not satisfied with the answer that klables are not objects, according to our conception of object, since he asks what is so sacred about *that* conception. It is at this point that philosophers often reach for transcendental arguments, but we shall content ourselves with a more modest response. Our ordinary concept of object does not admit klables. So if it is true that the circumscription of possibilities for entities resultant upon the formation of the conception of what it is to be a particular thing, the conception which we have identified as the notion grasp of which is required to make the step from *de dicto* to *de re* modality, if it is true that that conception stands to a thing's possibilities as the spatio-temporal continuity conception stands to a thing's past and future, then we have a relative justification of these modal concepts, in that putative alternatives would lead to analogously gerrymandered entities. Of course, there may be a whole alternative scheme of Goodmanesque concepts under which the gerrymandered entities would be appropriate, but in our scheme, it is notable that the rule for reidentifying a particular type of gerrymandered entity has a highly non-intrinsic nature. For example, to identify this klable at a later time, the question of exactly what time it

[24] Shoemaker [1979, pp. 336–9].

is at which the identification is to be made enters essentially, for it is crucial which side of noon the time lies on. Thus the intrinsicness of the rule is important to the distinction between the genuine and the gerrymandered, and in that case the essentialism we have been concerned to defend is maximally appropriate for fixing the boundaries of the possible for real rather than gerrymandered objects, given its relationship to intrinsicness which we have uncovered.[25,26]

[25] At the beginning of this chapter, the epistemological component of a justification of modal knowledge was mentioned. An attractive feature of our account of modal knowledge is that it does not render its possession mysterious in virtue of some *sui generis* inaccessibility of the facts; however, such knowledge will inherit the problems surrounding the general notion of *a priori* knowledge. A proper treatment of this notion could be given only in the context of a full theory of knowledge, which will not be attempted here. But the problem to be addressed is this. Granted that the point of drawing a distinction between knowledge and merely true belief is to differentiate reliable from unreliable methods of belief acquisition, so that knowledge is acquired only when it is *no accident* that in the circumstances a belief which is true is acquired, what we want of a method of acquiring *a priori* knowledge is that it reliably extract from mastery of a concept the principles or rules to which that mastery conforms. If concept-mastery is a type of *knowledge-how*, the method of acquiring the appropriate *knowledge-that* may be no different in its workings from general procedures for recovering principles underlying performances; the *a priori* status of the principles would be a consequence of the subject-matter the method is applied to, its status as knowledge a consequence of the reliability of the method in the circumstances of its application.

[26] The theory of *de re* necessity developed here is relevant to the question of which modal logic is the correct logic for broadly logical possibility and necessity, since it throws in doubt the coherence of the idea of a world accessible to some worlds and inaccessible to others. A set of possible worlds is a model of a putative modal reality, and we can say that such a model is *admissible* provided all *a priori* conceptual truths hold at every world. Someone tempted to speculate about 'alternative conceptual schemes' is therefore speculating about inadmissible models, not inaccessible worlds. Of the various admissible models, only one is the 'right' model, and we need *a posteriori* information to determine which it is, e.g. given representatives of organisms, we need to know the actual facts about the biological relationships amongst those organisms to distinguish the right model from one in which the relationship of being a propagule is rigid but there are impossible instances of that relation. Non-admissible models are therefore impossible, speaking in the broadly logical sense, since they contain impossible worlds, though perhaps only *a posteriori* impossible worlds. Now consider the suggestion that some world w in the right model is not accessible from the actual world, but is accessible from some world accessible from the actual world. Such a world w is contingently impossible, relative to w*. But in what could such impossibility consist? No *a priori* conceptual truth can fail at it, since it is then not a *possible* world at all (by definition, no such world is in the right, or even any admissible, model). Could some *a posteriori* necessary truth, necessary at w*, fail at w? Evidently not: the same *a priori* conceptual truths hold at every world, and any *a posteriori* truth T necessary at the actual world is so by being true at the actual world and by some conceptual truth's entailing that T's truth makes it necessary. Thus T holds at any world accessible to the actual world, so the same conceptual truth will make it necessary at such a world over again; hence we never reach a world where some actual impossibility is true. Since a world is accessible to the actual world provided everything true at it is actually possible, failure of transitivity of accessibility therefore never arises. Similar reasoning settles the question of symmetry, which means that S5 emerges as the correct system.

APPENDIX

Translation Schemes

1. *Sentential Modal Logic*

The language L_m to be translated is the language of ordinary propositional calculus with countably many sentence letters, supplemented with the sentential connectives '$\Box$' and '$\Diamond$', which are subject to the following clauses in the definition of well-formed-formula ('wff'):

($\Box$) If A is a wff, then $\Box$A is a wff.

($\Diamond$) If A is a wff, then $\Diamond$A is a wff.

The set of wffs (sentences) of L_m, Sent(L_m), is the least set containing all sentential letters and closed under the formation rules for the connectives.

If $\sigma \in$ Sent(L_m), the translation of σ, Trans[σ], is the *relativization* of σ to the actual world w*:

Definition: Trans[σ] = Rel[σ, w*].

The language of the sentences in the range of Trans is the language L_q; for each sentence letter π of L_m, L_q contains a monadic predicate $\pi(\xi)$. In addition, L_q contains the truth-functional connectives of L_m, quantifiers, countably many variables, and the individual constant 'w*'.

We give a recursive definition of the relativization function for an arbitrary world, so that it can be applied to any L_m sentence to yield an L_q sentence through step-by-step application of the clauses of the definition. In the clauses below, corner quotes are performing their usual function of selective quotation; thus, for example, for any L_m formula A, '$\ulcorner$—[Rel(A, w)]$\urcorner$' stands for the formula which consists in the symbol '—' followed by the expression which is the relativization of A to w (not: the relativization of 'A' to w). 'Rel' is a functor which, on substitution of an L_m sentence for the metalinguistic variable, forms a term for an L_q formula. The recursive definition of relativization is:

(1) If A is a sentence letter of L_m, Rel(A, w) = ⌜A(w)⌝.

(2) Rel(⌜–A⌝, w) = ⌜–[Rel(A, w)]⌝.

(3) Rel(⌜A & B⌝, w) = ⌜Rel(A, w) & Rel(B, w)⌝.

(4) Rel(⌜A v B⌝, w) = ⌜Rel(A, w) v Rel(B, w)⌝.

(5) Rel(⌜A → B⌝, w) = ⌜Rel(A, w) → Rel(B, w)⌝.

(6) Rel(⌜□A⌝, w) = ⌜(∀u)[Rel(A, u)]⌝.

(7) Rel(⌜◇A⌝, w) = ⌜(∃u)[Rel(A, u]⌝.

We use 'w', 'u', 'v', etc., as metalinguistic variables ranging over variables of L_q and also as variables of L_q themselves. Clearly, in (6) and (7) they are metalinguistic, and in particular applications of (6) and (7) their instances are to be chosen to avoid scope clashes.

To illustrate how the translation scheme works, we translate the L sentence '□(P → (Q & ◇R))'. By the definition, we have

$$\text{Trans}[\Box(P \rightarrow (Q \,\&\, \Diamond R))] = \text{Rel}[\Box(P \rightarrow (Q \,\&\, \Diamond R)), w^*]$$

which by (6) is

$$(\forall w)(\text{Rel}[P \rightarrow (Q \,\&\, \Diamond R), w])$$

which by (5) is

$$(\forall w)(\text{Rel}[P, w] \rightarrow \text{Rel}[Q \,\&\, \Diamond R, w])$$

which by (1) and (3) is

$$(\forall w)(P(w) \rightarrow \text{Rel}[Q, w] \,\&\, \text{Rel}[\Diamond R, w])$$

which by (1) and (7) is

$$(\forall w)(P(w) \rightarrow (Q(w) \,\&\, (\exists u)R(u))).$$

Modulo choices of L_q variables at the second and last line, this last displayed formula is the correct translation into L_q of the original L_m sentence.

2. *Quantified S5*

L_m is now a first-order language supplemented with '□' and '◇', which are subject to the same clauses in the definition of wff as before. Trans is defined for formulae, and if $\phi \in \text{Form}(L_m)$,

$$\text{Trans}[\phi] = \text{Rel}[\phi, w^*].$$

L_w, into which we translate L_m sentences, is a two-sorted first-order language which contains all the connectives of L_m except the modal ones, all the individual constants and variables of L_m, countably many variables of a new sort (sort 1, or sort W, i.e. world variables), an individual constant of the new sort ('w*'), a two-place predicate $E(\xi, \zeta)$ which takes an object-variable in its first place and a world variable in its second (i.e. its category is $\langle 2, 1\rangle$—this stipulation is unnecessary if L_m already contains a one-place existence predicate) and for each n-place predicate F of L_m an n+1-place predicate $F(\xi_1 \ldots \xi_n, \zeta)$, where ζ is a world variable. Rel is defined for the truth-functional connectives as before, and we have the following new clauses:

(8) If $F(t_1 \ldots t_n)$ is an atomic formula (t_i either a free variable or a constant), then

$$\mathrm{Rel}[F(t_1 \ldots t_n), w] = \ulcorner F(t_1 \ldots t_n, w)\urcorner.$$

(9) $\mathrm{Rel}[\ulcorner t_1 = t_2\urcorner, w] = \ulcorner t_1 = t_2\urcorner$.

(10) $\mathrm{Rel}[\ulcorner(\forall x)\phi(x)\urcorner, w] = \ulcorner(\forall x)(E(x, w) \rightarrow \mathrm{Rel}[\ulcorner\phi(x)\urcorner, w])\urcorner$.

(11) $\mathrm{Rel}[\ulcorner(\exists x)\phi(x)\urcorner, w] = \ulcorner(\exists x)(E(x, w)\ \&\ \mathrm{Rel}[\ulcorner\phi(x)\urcorner, w])\urcorner$.

(12) $\mathrm{Rel}[\ulcorner\Diamond A\urcorner, w] = \ulcorner(\exists u)(\mathrm{Rel}(A, u))\urcorner$.

(13) $\mathrm{Rel}[\ulcorner\Box A\urcorner, w] = \ulcorner(\forall u)(\mathrm{Rel}(A, u))\urcorner$.

In these clauses, 'x' is a metalinguistic variable ranging over the individual variables of L_w. To illustrate their application:

$$\begin{aligned}\mathrm{Trans}[\Diamond(\exists x)Fx \rightarrow (\exists x)\Diamond Fx] &= \mathrm{Rel}[(\Diamond(\exists x)Fx \rightarrow (\exists x)\Diamond Fx), w^*]\\ &= (\exists w)\mathrm{Rel}[(\exists x)Fx, w] \rightarrow (\exists x)(E(x, w^*)\ \&\ \mathrm{Rel}[Fx, w^*])\\ &= (\exists w)(\exists x)(E(x, w)\ \&\ \mathrm{Rel}[Fx, w]) \rightarrow\\ &\qquad (\exists x)(E(x, w^*)\ \&\ (\exists w)\mathrm{Rel}[Fx, w])\\ &= (\exists w)(\exists x)(E(x, w)\ \&\ Fxw) \rightarrow (\exists x)(E(x, w^*)\ \&\ (\exists w)Fxw).\end{aligned}$$

Modulo choice of variables, the last displayed formula is the correct translation of formula (16) in §1 of Chapter 2.

3. *Counterpart-Theoretic S5*

To translate sentences of L_m (as in (2)) into the language L_c of counterpart theory, where L_c is just like L_w above except that it also has a three-place predicate $C(\xi_1, \xi_2, \zeta)$, the translation scheme of (2) can be used except for its modal operator clauses (12) and (13). We will give here the replacement clauses, and also the clauses for the un-indexed 'actually' operator. It is a reader's exercise to provide further clauses for more complex 'actually' operators.

For each modal operator there are two cases, according to whether the formula it governs does or does not contain occurrences of terms within the scope of some other modal operator in that formula; that is, in one case, some term-occurrences in A are 'immediately' within the scope of the governing modal operator, while, in the other, there are modal operators in A such that each term-occurrence in A is governed by at least one of those operators. A term is either a free variable or an individual constant.

If all term-occurrences in A are governed by modal operators in A, then we have the clauses (12) and (13) again:

(12) $\mathrm{Rel}[\ulcorner \Diamond A \urcorner, w] = \ulcorner(\exists u)(\mathrm{Rel}[A, u])\urcorner$.

(13) $\mathrm{Rel}[\ulcorner \Box A \urcorner, w] = \ulcorner(\forall u)(\mathrm{Rel}[\forall, u])\urcorner$.

However, if $t_1 \ldots t_n$ are n occurrences of constants or free variables not within the scope of some modal operator in A, then we have these clauses:

(14) $\mathrm{Rel}[\ulcorner \Diamond A \urcorner, w] = (\exists u)(\exists x_1) \ldots (\exists x_n)(Cx_1t_1u \mathbin{\&} \ldots \mathbin{\&} Cx_nt_nu \mathbin{\&} \mathrm{Rel}[A(x_i/t_i), u])$.

(15) $\mathrm{Rel}[\ulcorner \Box A \urcorner, w] = (\forall u)(\forall x_1) \ldots (\forall x_n)(Cx_1t_1u \mathbin{\&} \ldots \mathbin{\&} Cx_nt_nu \rightarrow \mathrm{Rel}[A(x_i/t_i), u])$.

The point of distinguishing the two cases for each modal operator is to ensure that the propositional behaviour of the modal operators is unaffected by the counterpart relation. For instance, in S5 we want '$\Box$Fa' and '$\Box\Box$Fa' to be equivalent even if the counterpart relation is not transitive (even if counterparts of a's counterparts are not always counterparts of a). Since the first occurrence of '$\Box$' in '$\Box\Box$Fa' will be dealt with by clause (13) it will not introduce the counterpart predicate, so questions about counterparts of a's counterparts will not arise.

Similar comments apply to the 'actually' operator, so we have two clauses for it, dealing with the respective cases:

(16) $\mathrm{Rel}[\ulcorner A\phi \urcorner, w] = \mathrm{Rel}[\phi, w^*]$.

(17) $\text{Rel}[\ulcorner \mathsf{A}\phi \urcorner, w] = \ulcorner(\exists x_1)\ldots(\exists x_n)(Cx_1t_1w^* \,\&\, \ldots Cx_nt_nw^*$
$\&\ \text{Rel}[\phi(x_i/t_i), w^*])$.

The reader may use the above scheme to confirm that the problematic formula '$(\forall x)\Box(\exists y)(x = y)$', which Lewis's scheme translates into a theorem of counterpart theory, is translated by our scheme into the following non-theorem:

$$(\forall x)(E(x, w^*) \rightarrow (\forall w)(\forall z)(Czxw \rightarrow (\exists y)(E(y, w) \,\&\, z = y))).$$

Note 24 to Chapter 7 contains a translation-scheme clause for the counterfactual conditional '$\Box\!\!\rightarrow$'.

4. *Reverse translation*

The reverse-translation procedure which maps some L_w sentences back into L_m (with singly-indexed 'actually' operators) is just the inverse of the procedure in 2 above. The procedure may be specified as a sequence of machine-executable instructions; given an appropriate L_w sentence A we:

(i) replace each world quantifier in A with an indexed '$\Box$' or '$\Diamond$', without repetition of indexes;
(ii) along with each such replacement, eliminate the world-variable freed by the deletion of the quantifier in the rest of the formula, placing a co-indexed 'actually' operator in front of each atomic formula from which an occurrence of the variable is deleted;
(iii) eliminate 'w^*' from each atomic formula in which it occurs and prefix the formula with 'A';
(iv) erase redundant indexes, 'actually' operators, and occurrences of the existence predicate.

'Redundant' needs some explanation. An index on a '$\Box$' or a '$\Diamond$' is redundant if there is no occurrence of 'A_i' within the range of some other '$\Box$' or '$\Diamond$' (indexed or not) itself within the scope of the indexed modal operator. An index i on 'A' is redundant if i also indexes some modal operator and no other modal operator is between it and that operator. 'A' is redundant if it is not within the scope of any '$\Box$', '$\Diamond$', or 'A_i'. $\ulcorner E(x) \urcorner$ is redundant if x is bound by a quantifier such that no '$\Box$', '$\Diamond$' or 'A' (indexed or unindexed) occurs between it and the quantifier.

5. *Possibilist Quantifiers*

Our decision to interpret the objectual quantifiers of modal language in the *actualist* manner, on which a quantified sentence is evaluated at a world w by restricting the scope of the relevant quantifier to d(w), is not the only possible treatment. We might instead have given the following clauses:

(P∀) $w \vDash (\forall v)Av$ iff for every a in D, $w \vDash Av[\underline{a}/v]$.

(P∃) $w \vDash (\exists v)Av$ iff for some a in D, $w \vDash Av[\underline{a}/v]$.

With such clauses, the objectual quantifiers are sometimes called 'outer' or 'possibilist' quantifiers, since at any world they range over the domain of all possible objects of the model. An obvious question to ask is whether anything turns on the choice we make between these readings of '∀' and '∃', a question we shall answer in the negative: we show below that any sentence containing possibilist quantifiers is logically equivalent to some sentence with actualist quantifiers, and conversely. In terms of a contrast between evaluation clauses, this means that the proposition expressed by a given sentence when its quantifiers are interpreted as possibilist can also be expressed by a sentence whose quantifiers are interpreted as actualist, and vice versa. However, we can give a more manageable formulation of this claim if we use special symbols for the possibilist quantifiers, and establish the result by giving a translation scheme taking sentences with these symbols into sentences without. We will use 'Π' for the possibilist universal quantifier, with the evaluation condition given on the right hand side of (P∀), and 'Σ' for the possibilist existential quantifier, with the evaluation condition on the right hand side of (P∃); since the quantifiers are interdefinable, we focus attention on 'Π'.

We shall first establish the following: if L is a first-order modal language with standard propositional connectives, the modal operators '□' and '◇', and both pairs of quantifiers, and if L* is a first-order modal language with standard propositional connectives, actualist quantifiers only, the modal operators '□' and '◇' (indexed or unindexed), and the 'actually' operator (again indexed or unindexed), then for each sentence σ of L there is a sentence σ^* of L* such that σ and σ^* hold in the same quantified S5 models.

Clearly, all that we need to find are recursive translation clauses for formulae of the form

$$(\Pi v)Av$$

and

$$(\Sigma v)Av.$$

However, although '$(\Pi x)Fx$' has the sense 'for all possible x, x is F', it would be incorrect to propose that this formula be translated either by '$\Box(\forall x)Fx$' or by '$\Box(\forall x)\Box Fx$'. For suppose

(18) $w \models (\Pi x)Fx.$

Then this means that, at w, all members of D are assigned to the extension of F. But it does not follow that for every w' in W, all members of $d(w')$ are assigned to the extension of F; indeed, the only w' for which we can be sure this condition holds is w itself. Hence we cannot conclude:

(19) $w \models \Box(\forall x)Fx$

nor, *a fortiori*,

(20) $w \models \Box(\forall x)\Box Fx.$

Conversely, (19) does not entail (18), since (19) does not say, of the non-existents at w, that they are F *at w*; (20), however, does imply (18), since it says that every possible object is in the extension of F at *every* world.

The problem is to combine a quantification over all possible objects x, '$\Box(\forall x)$', with a claim about how each such x is at a *given* world, the same world for each x. In case this world is the actual world, i.e. in case the assertion is just 'all possible x are F', we can use the 'actually' operator to fix the world at which '$F\underline{a}$' is evaluated, for each possible object a. Thus the following is a logical truth (true at the actual world of every model):

(21) $(\Pi x)Fx \leftrightarrow \Box(\forall x)\mathsf{A}(Fx).$

With more complex L sentences, the possibilist quantifiers may be inside the scope of a '$\Box$' or a '$\Diamond$', so what is required is an L* translation with indexed 'actually' operators such that, when a claim about all possibles is evaluated at a world introduced by a modal operator in the L sentence, an indexed 'actually' operator takes us to that same world at the corresponding stage of the evaluation of its L* translation. For instance, the following is also logically true:

(22) $\Diamond(\Pi x)Fx \leftrightarrow \Diamond_1\Box(\forall x)\mathsf{A}_1(Fx).$

The translation procedure to be described simply generalizes the idea underlying (22).

If σ is a sentence of L with n occurrences of modal operators, let σ' be the L* sentence which results from σ by indexing the i'th occurrence of a modal operator in σ with the numeral i, $1 \leqslant i \leqslant n$. Then for each occurrence of $\ulcorner(\Pi v)\urcorner$ or $\ulcorner(\Sigma v)\urcorner$ which is immediately within the scope of the i'th modal operator in σ, for $1 \leqslant i \leqslant n$ replace:

(i) that occurrence of $\ulcorner(\Pi v)\urcorner$ in σ' with $\ulcorner\Box(\forall v)A_i\urcorner$

and

(ii) that occurrence of $\ulcorner(\Sigma v)\urcorner$ in σ' with $\ulcorner\Diamond(\exists v)A_i\urcorner$.

while if an occurrence of $\ulcorner(\Pi v)\urcorner$ or $\ulcorner(\Sigma v)\urcorner$ in σ is not within the scope of any modal operator in σ then replace it with $\ulcorner\Box(\forall v)A\urcorner$ or $\ulcorner\Diamond(\exists v)A\urcorner$, as appropriate, in σ'. It is easy to see, and trivial to prove, that the sentence σ^* which results from σ by this procedure is logically equivalent to σ.

It should be clear that, given a language with standard modal operators and possibilist quantifiers only, any sentence of the actualist language (without 'actually' operators) can be expressed, since one can obtain the effect of $\ulcorner(\forall v)A\urcorner$ in such a language by $\ulcorner(\Pi v)(E(v) \rightarrow A)\urcorner$. Thus a converse to the above result obtains, which can be strengthened by adding 'actually' operators to *both* languages and qualifying the description of the construction of σ' to avoid clashes of indexes. It therefore follows that there is neither loss nor gain in expressive power in our choice of the actualist reading of the quantifiers for S5. From the philosophical point of view, however, it is natural to think of this result as an eliminability result for possibilist quantifiers: sentences of the possibilist language are thereby shown to be true in virtue of sentences of the actualist language.

This result holds generally only when the actualist language has indexed 'actually' operators, for without such operators the non-equivalence of (18) with either (19) or (20) blocks an elimination. However, something can be accomplished even without the indexed operators, since we might choose to lay down axioms which prevent the sort of situation which refutes the equivalence of (18) with (19) or (20) from arising. That is, although there is no general eliminability with respect to standard modal language, there may be a reasonable notion of 'eliminability in the theory T', where T is a theory in standard modal language whose axioms have the desired effect. Let us consider (20): the strength of (20) is in excess of that of (18) because (20) requires that for each c in D, 'F$\underline{c}$' be true at every world, rather than simply at w. Hence a rather crude criterion for eliminability within a theory is that if for every formula ϕ with n free variables

$$T \vdash \Box(\forall x_1)\Box \ldots \Box(\forall x_n)\Box(\phi \rightarrow \Box\phi)$$

then T permits elimination of $\ulcorner(\Pi v)\urcorner$ by the definition $\ulcorner\Box(\forall v)\Box\urcorner$.

But more interesting criteria are possible. One example, due to Kit Fine [1981b, 192–3], can be motivated by consideration of a modal set theory such as MST (see §3 of Chapter 5). In modal set theory, every pure set is a necessary existent, necessarily a set, and stands in the same membership relations to the same pure sets at each world (this is consistent with there being independent sentences in ZF: it is just that distinct models of ZF could not figure as distinct worlds within a *single* model of MST). Thus if a model for MST contains two worlds u and v which are not isomorphic, this must trace to differences between these worlds in terms of their actual individuals; in fact, given the axioms of MST, the difference would have to be in the cardinality of the set of individuals at each, but, certainly, if the individuals of two worlds were identically the same individuals, then the worlds could not differ. In this case, then, 'all possible individuals are F' will hold at a world w iff 'all possible individuals are F at every world with the same individuals as this one' holds at w, and this latter claim is quasi-formalizable as '$\Box(\forall x)\Box$(if the individuals are the same as in w, then Fx)': it is equivalent to 'all possible individuals are F', although its second '$\Box$' quantifies over worlds other than w, because these other worlds are just like w (in modal set theory). To obtain a first-order definition from this, we have to eliminate the reference to w by formalizing the idea of some world having the same existent individuals as a given world, which cannot be done just with quantifiers ranging over individuals (unless we reintroduce the indexed operators or their like). But since the existent individuals of a world form a set, and a set is an entity, we can say that the individuals of some world u are the same as those of a given world v by saying that the set of existent individuals at v exists at u and is also the set of existent individuals there. Since a set cannot change its members through worlds, this means u and v have the same domain. Thus, in modal set theory, we have the following definition:

Definition: $(\Pi x)A =_{df} (\exists y)(Ry \,\&\, \Box(\forall x)\Box(Ry \rightarrow A))$

where 'Ry' abbreviates

$$Sy \,\&\, E(y) \,\&\, (\forall z)(Iz \leftrightarrow z \in y)$$

and 'y' is not free in A ('S' means 'set' and 'I', 'individual').

From the set-theoretic case we can isolate the general features on which the correctness of this definition relies. The first feature is that at every world there is a set of existing individuals, since if this were not true, then if w is a world at which no object satisfies 'R', the *definiens* as it stands is automatically false at w, even if A holds of all possibles at w, while if '$(\exists y)(Ry\,\&$' were altered to '$(\forall y)(Ry \rightarrow$', the *definiens* would be automatically true for all A. The second feature is that the expressible facts about any possible individuals are the same in any two worlds which share the same set of individuals, or, more generally, which share the same satisfier of 'R' (the intuitive meaning of 'R' is irrelevant to the formal correctness of the definition). Thus we arrive at Fine's general criterion for any theory T of a modal language with just '$\Box$' and '$\Diamond$' as modal operators to permit elimination of possibilist quantifiers: there must be a formula 'R(v)' with one free variable such that

(i) $T \vdash \Box(\exists x)Rx$

and such that for any atomic formula ϕ whose free variables are $y_1 \ldots y_n$,

(ii) $T \vdash \Box(\forall x)\Box(\forall y_1) \ldots \Box(Tx \,\&\, \phi \rightarrow \Box(Tx \rightarrow \phi))$.

In such a theory, possibilist quantifiers may be eliminated from sentences containing them in accordance with the definition, for the resulting sentences will be provably equivalent to the original ones in the theory T.

Clearly, Fine's criterion will be satisfied only by theories about rather special sorts of entities: besides sets, propositions and possible worlds themselves, for example. Moreover, there does not seem to be any alternative with a significantly wider range of application. From these facts, therefore, we may infer another justification for introducing the indexed 'actually' operators into modal language; after all, possibilist quantifiers are perfectly coherent, yet it does seem plausible to hold that we understand them in terms of actualist quantifiers and modalities. And what we have just seen is that, in general, an actualist formulation of a possibilist statement needs the special operators.

Bibliography

Adams, R. M. [1974], 'Theories of Actuality', *Nous* 8 (1974), 211–31; Loux, M. J. (ed.) (1979), 190–209.

— [1979], 'Primitive Thisness and Primitive Identity', *The Journal of Philosophy* 76 (1979), 5–26.

— [1981], 'Actualism and Thisness', *Synthese* 49 (1981), 3–41.

Aleksandrov, A. D., Kolmogorov, A. N., and Lavrent'ev, M. A. (eds.) [1983], *Mathematics: Its Content, Methods, and Meaning* (3 vols.), The M.I.T. Press, 1963.

Almog, J. [1981], 'Dthis and Dthat: Indexicality Goes Beyond That', *Philosophical Studies* 39 (1981), 347–81.

Aristotle [1928], *The Works of Aristotle*, Vol. I (trans. Edgehill, E. M.), Oxford University Press, 1971.

Averill, E. A. [1982], 'The Primary-Secondary Quality Distinction', *The Philosophical Review* 91 (1982), 343–61.

Barwise, J. [1977], *The Handbook of Mathematical Logic*, North Holland, 1977.

Benacerraf, P. [1965], 'What Numbers Could Not Be', *The Philosophical Review* 74 (1965), 47–73.

Bencivenga, E. [1976] 'Set Theory and Free Logic', *The Journal of Philosophical Logic* 5 (1976), 1–15.

van Benthem, J. [1978], 'Two Simple Incomplete Modal Logics', *Theoria* 44 (1978), 25–37.

— [1979], 'Syntactic Aspects of Modal Incompleteness Theorems', *Theoria* 45, 63–77.

Blok, W. [1980], 'The Lattice of Modal Logics: an Algebraic Investigation' ,*The Journal of Symbolic Logic* 45 (1980), 221–36.

Boolos, G. [1971], 'The Iterative Conception of Set', *The Journal of Philosophy* 68 (1971), 215–31.

— [1979], *The Unprovability of Consistency*, Cambridge University Press, 1979.

Butts, R., and Hintikka, J. (eds.) [1977], *Logic, Foundations of Mathematics and Computability Theory*, Reidel, 1977.

Chandler, H. [1976], 'Plantinga and the Contingently Possible', *Analysis* 35 (1976), 106–9.

Chellas, B. F. [1980], *Modal Logic*, Cambridge University Press, 1980.

Chisholm, R. [1968] 'Identity Through Possible Worlds: Some Questions', *Nous* 1 (1968), 1–8.

— [1970], 'Identity Through Time', in Kiefer, H. E. and Munitz, M. (eds.) (1970), 163–82.

Davidson, D. [1969], 'True to the Facts', *The Journal of Philosophy* 66 (1969), 748–764.

Davidson, D., and Harman G. (eds.) [1972], *Semantics of Natural Language*, Reidel, 1972.

Davies, M. K. [1978], 'Weak Necessity and Truth Theories', *The Journal of Philosophical Logic* 7 (1978), 415–39.

— [1981], *Meaning, Quantification, Necessity: Themes in Philosophical Logic*, Routledge and Kegan Paul, 1971.

— [1983], 'Actuality and Context-Dependence II', *Analysis* 43 (1983), 128–33.

— and Humberstone, L. [1980], 'Two Notions of Necessity', *Philosophical studies* 38 (1980), 1–30.

Devlin, K. [1979], *Fundamentals of Contemporary Set Theory*, Springer Verlag, 1979.

Dugundji, J. [1940], 'Note on a Property of Matrices for Lewis and Langford's Calculi of Propositions', *The Journal of Symbolic Logic* 5 (1940), 151–1.

Dummett, M. [1974], *Frege: Philosophy of Language*, Duckworth, 1973.

— [1975a], 'Wang's Paradox', *Synthese* 30 (1975), 301–24; Dummett (1978), 248–68.

— [1975b], 'The Philosophical Basis of Intuitionistic Logic', in Rose, H. E., *et al.* (eds.), (1975), 5–40; Dummett (1978), 215–47.

— [1977], *Elements of Intuitionism*, Oxford University Press, 1977.

— [1978], *Truth and Other Enigmas*, Duckworth, 1978.

— [1982], 'Realism', *Synthese* 52 (1982), 55–112.

Dupré, J. [1981], 'Natural Kinds and Biological Taxa', *The Philosophical Review* 90 (1981), 66–90.

Ede, D. A. [1978], *An Introduction to Developmental Biology*, Blackie 1978.

Enderton, H. B. [1972], *A Mathematical Introduction To Logic*, Academic Press, 1972.

Evans, G. [1979], 'Reference and Contingency', *The Monist* 62 (1979), 161–84.

— [1982], *The Varieties of Reference*, Oxford University Press, 1982.

Field, H. [1977], 'Logic, Meaning and Conceptual Role', *The Journal of Philosophy* 74 (1977), 379–409.

— [1980], *Science Without Numbers*, Blackwell, 1980.

Fine, K. [1975], 'Vagueness, Truth and Logic', *Synthese* 30 (1975), 265–300.

— [1977], 'Properties, Propositions and Sets', *The Journal of Philosophical Logic* 6 (1977), 135–91.

— [1978a], 'Model Theory for Modal Logic Part I: the *De Re/De Dicto* Distinction', *The Journal of Philosophical Logic* 7 (1978), 125–56.

— [1978b], 'Model Theory for Modal Logic Part II: The Elimination of the *De Re*', *The Journal of Philosophical Logic* 7 (1978), 277–306.

— [1981a], 'Model Theory for Modal Logic Part III: Existence and Predication', *The Journal of Philosophical Logic* 10 (1981), 293–307.

— [1981b], 'First-Order Modal Theories I—Sets', *Nous* 15 (1981), 177–205.

Forbes, G. [1980], 'Origin and Identity', *Philosophical Studies* 37 (1980), 353–62.

— [1981], 'On The Philosophical Basis of Essentialist Theories', *The Journal of Philosophical Logic* 10 (1981), 73–99.

— [1983], 'Actuality and Context-Dependence I', *Analysis* 43 (1983), 123–8.

— [1984a], 'Scepticism and Semantic Knowledge', *Proceedings of the Aristotelian Society* (1983/4), 223–37.

— [1984b], 'Places as Possibilities of Location', forthcoming.

French, P. A., Uehling, T. E., and Wettstein, H. K. (eds.) [1979], *Studies in Metaphysics; Midwest Studies in Philosophy Volume 4*, University of Minnesota Press, 1979.

Gentzen, G. [1969], *The Collected Papers of Gerhard Gentzen*, North Holland 1969.

Glymour, C. [1980], *Theory and Evidence*, Princeton University Press, 1980.

Goguen, J. [1969], 'The Logic of Inexact Concepts', *Synthese* 19 (1969), 325–73.

Goldblatt, R. I. [1975], 'First-Order Definability in Modal Logic', *The Journal of Symbolic Logic* 40 (1975), 35–40.

— [1976a], 'Metamathematics of Modal Logic Part I', *Reports on Mathematical Logic* 5 (1976), 41–78.

— [1976b], 'Metamathematics of Modal Logic Part II', *Reports on Mathematical Logic* 7 (1976), 21–52.

Grandy, R. [1982], review article, *The Journal of Symbolic Logic* 47 (1982), 689–94.

Guenther, F., and Guenther-Reutter, M. (eds.) [1978], *Meaning and Translation*, Duckworth, 1978.

Gupta, A. [1980], *The Logic of Common Nouns*, Yale University Press, 1980.

Hazen, A. [1976], 'Expressive Incompleteness in Modal Logic', *The Journal of Philosophical Logic* 5 (1976), 25–46.

— [1979], 'Counterpart Theoretic Semantics for Modal Logic', *The Journal of Philosophy* 76, 319–38.

Healey, R. (ed.) [1981], *Reduction, Time and Reality*, Cambridge University Press, 1981.

Hintikka, J. [1975], *The Intentions of Intentionality and Other New Models for Modalities*, Reidel, 1975.

Hirsch, E. [1971], 'Essence and Identity', in Munitz, M. (ed.) (1971), 31–51.

Hodges, W. [1977], *Logic*, Penguin Books, 1977.

Hughes, G. E., and Cresswell, M. J. [1968], *An Introduction to Modal Logic*, Methuen 1968.

Humberstone, L. [1981], 'From Worlds to Possibilities', *The Journal of Philosophical Logic* 10, 313–39.

Kalish, D., Montague, R., and Mar, G. [1980], *Logic: Techniques of Formal Reasoning*, Harcourt Brace Jovanovich (2nd edn), 1980.

Kaplan, D. [1977], Demonstratives, unpublished manuscript.

— [1979], 'Transworld Heirlines', in Loux, M. J. (1979), 88–109.

Kearns, J. [1981], 'Modal Semantics Without Possible Worlds', *The Journal of Symbolic Logic* 46 (1981), 77–82.

Kiefer, H. E., and Munitz, M. (eds.) [1970], *Language, Belief and Metaphysics*, State University of New York Press, 1970.

Kripke, S. [1963], 'Semantical Considerations on Modal Logic', *Acta Philosophica Fennica* 16 (1963), 83–94; Linsky, L. (ed.) (1971), 63–72.

— [1971], 'Identity and Necssity', in Munitz, M. (ed.) (1971), 135–64.

— [1972], 'Naming and Necssity', in Davidson, D., and Harman, G. (1972), 252–355.

Lemmon, E. J., and Scott, D. S. [1977], *The 'Lemmon Notes': An Introduction to Modal Logic, American Philosophical Quarterly Monograph Series*, no. 10, Blackwell, 1977.

Lewis, D. [1968], 'Counterpart Theory and Quantified Modal Logic', *The Journal of Philosophy* 65 (1968), 113–26; Loux, M. J. (1979), 110–28.

— [1970], 'Anselm and Actuality', *Nous* 4 (1970), 175–88.

— [1973a], *Counterfactuals*, Blackwell, 1973.

— [1973b], 'Causation', *The Journal of Philosphy* 70 (1973), 556–7.

Linsky, L. (ed.) [1971], *Reference and Modality*, Oxford University Press, 1971.

Lombard, L. B. [1979], 'Events', *The Canadian Journal of Philosophy* 9 (1979), 425–60.

— [1981], 'Events and Their Subjects', *The Pacific Philosophical Quarterly* 62 (1981), 138–47.

— [1982], 'Events and the Essentiality of Time', *The Canadian Journal of Philosophy* 12 (1982), 1–17.

Loux, M. J. (ed.) [1979], *The Possible and The Actual*, Cornell University Press, 1979.

Mackie, J. L. [1974], '*De* What *Re* is *De Re* Modality?', *The Journal of Philosophy* 71 (1974), 551–61.

Marcus, R. B. [1962], 'Interpreting Quantification', *Inquiry* 5 (1962), 252–9.

Maynard Smith, J. [1979], *The Theory of Evolution*, Penguin Books (3rd edn), 1975.

McGinn, C. [1976], 'On The Necessity of Origin', *The Journal of Philosophy* 73 (1976), 127–35.

— [1981], 'Modal Reality', in Healey, R. (ed.) (1981), 143–87.

Mondadori, F., and Morton, A. [1976], 'Modal Realism: The Poisoned Pawn', *The Philosophical Review* 85 (1976), 3–20; Loux, M. (ed.) (1979), 235–52.

Moravcsik, J. M. E. (ed.) [1967], *Aristotle; A Collection of Critical Essays*, MacMillan, 1968.

Munitz, M. (ed.) [1971], *Identity and Individuation*, New York University Press, 1971.

Nozick, R. [1981], *Philosophical Explanations*, Harvard Belknap, 1981.

Parfit, D. [1971], 'Personal Identity', *The Philosophical Review* 80 (1971), 3-27.

Parsons, C. [1971], 'A Plea for Substitutional Quantification', *The Journal of Philosophy* 68 (1971), 231-7.

— [1977], 'What is the Iterative Conception of Set?', in Butts, R., and Hintikka, J. (eds.) (1977), 335-67.

Peacocke, C. [1978], 'Necessity and Truth Theories', *The Journal of Philosophical Logic* 7 (1978), 473-500.

— [1979], *Holistic Explanation*, Oxford University Press, 1979.

— [1981], 'Are Vague Predicates Incoherent?', *Synthese* 46 (1981), 121-41.

— [1982], 'Causal Modalities and Realism', in Platts, M. (ed.) (1982), 41-68.

— [1983a], *Sense and Content*, Oxford University Press, 1983.

— [1983b], 'Knowledge', unpublished typescript.

Pears, D. (ed.) [1972], *Russell's Logical Atomism*, Fontana, 1972.

Plantinga, A. [1974], *The Nature of Necessity*, Oxford University Press, 1974.

Platts, M. (ed.) [1980], *Reference, Truth and Reality*, Routledge and Kegan Paul, 1980.

Prawitz, D. [1971], 'Towards a Foundation of a General Proof Theory', in Suppes, P., *et al.* (eds.) (1973), 225-50.

Prior, A. N. [1967], *Past, Present and Future*, Oxford University Press, 1967.

— and Fine, K. [1976], *Worlds, Times and Selves*, Duckworth, 1976.

Putnam, H. [1978], 'Meaning, Reference and Stereotypes', in Guenther, F. and Guenther-Reutter, M. (eds.) (1978), 61-81.

Quine, W. V. O. [1961], *From a Logical Point of View*, Harvard University Press, 1961.

— [1963], *Set Theory and Its Logic*, Harvard Belknap, 1963.

— [1966], *The Ways of Paradox*, Random House, 1966.

— [1970], *Philosophy of Logic*, Prentice-Hall, 1970.

— [1976], 'Worlds Away', *The Journal of Philosophy* 73 (1976), 859-863.

Rescher, N., and Urquart, A. I. F. [1971], *Temporal Logic*, Springer, 1971.

Russell, B. [1918], *The Philosophy of Logical Atomism*, in Pears, D. (ed.) (1972), 31-142.

Salmon, N. [1979], 'How Not to Derive Essentialism from the Theory of Reference', *The Journal of Philosophy* 76 (1979), 703-725.

— [1981], *Reference and Essence*, Princeton University Press, 1981.

Schiffer, S. [1978], 'The Basis of Reference', *Erkenntnis* 13 (1978), 171-206.

Schock, R. [1968], *Logics Without Existence Assumptions*, Almqvist and Wiksell, 1968.

Schoenman, R. [1967], *Bertrand Russell: Philosopher of the Century*, George Allen & Unwin, 1967.

Scott, D. [1967], 'Existence and Description in Formal Logic', in

Schoenman (ed.) (1976), 181–200.
— [1971], 'On Engendering an Illusion of Understanding', *The Journal of Philosophy* 68 (1971), 787–807.
Scott, D. *et al.* [1981], *Notes on the Formalization of Logic*, Oxford Subfaculty of Philosophy, 1981.
Sharvy, R. [1968], 'Why a Class Can't Change its Members', *Nous* 2 (1968), 303–14.
Shoemaker, S. [1979], 'Identity, Properties and Causality', in French *et al.* (eds.) (1979).
Smorynski, C. [1977], 'The Incompleteness Theorems', in Barwise (ed.) (1977).
Smullyan, A. F. [1948], 'Modality and Description', *The Journal of Symbolic Logic* 13 (1948), 31–7; Linsky (ed.) (1971), 35–43.
Stroud, B. [1977], *Hume*, Routledge and Kegan Paul, 1977.
Suppes, P. [1972], *Axiomatic Set Theory*, Dover, 1972.
Suppes, P., Henkin, L., and Tarski, A. (eds.) [1973], *Logic, Methodology and Philosophy of Science*, vol. IV, North Holland, 1973.
Tarski, A. [1956], *Logic, Semantics and Metamathematics*, Oxford University Press, 1956.
Taylor, R. [1974], *Metaphysics* (2nd edn), Prentice-Hall, 1974.
van Dalen, D. [1980], *Logic and Structure*, Springer-Verlag, 1980.
Wertheimer, R. [1971], 'Understanding the Abortion Argument', *Philosophy and Public Affairs* 1 (1971), 67–95.
Wiggins, D. [1980], *Sameness and Substance*, Blackwell, 1980.
Wright, C. [1975], 'On The Coherence of Vague Predicates', *Synthese* 30 (1975), 325–65.
— [1980], *Wittgenstein on The Foundations of Mathematics*, Duckworth, 1980.

Index

abortion 167 n.
accessibility 12, 237 n.; in B, 15; in S4, 16; in S5, 15; in T, 17; in tense logic, 42
actuality 34, 75; as operator, 77, 90 ff., 241; indexed, 91 ff., 112 ff.
Adams, R. 75 n., 77 n., 149 ff., 188, 248
Aleksandrov, A. 78 n., 248
Almog, J. 196 n., 248
Aristotle 2 n., 248
atheism 23 ff.
atomism 167 n.
Averill, E. 204 n., 248

Barcan formulae 28, 38
Benacerraf, P. 80 n., 248
Bencivenga, E. 121 n., 248
Blok, W. 248
Boolos, G. 17 n., 102 n., 248

causal necessity 217 ff.
Chellas, B. 17 n., 248
Chisholm, R. 130 n., 162, 248
Chisholm's Paradox 65 n., 162 ff., 167 n., 184
conceivability 222 ff., 231 ff.
conventionalism 219 ff., 221
Counterpart Theory 57 ff., 177 ff.
Crossworld Extensionality 110 ff., 112, 123 ff., 134, 146, 211 n.

Davies, M. vi, 30 n., 35 n., 75 n., 77 n., 90 n., 205 n., 231 n., 249
descriptions 72 ff.
Devlin, K. 103 n., 249
Dugundji, J. 4 n., 249
Dummett, M. vi, 75 n., 83, 84 n., 85, 89, 94 n., 166, 167 n., 169 n., 201 n., 217, 225 n., 232 n., 249
Dupré, J. 194 n., 249

Ede, D. 160 n., 249
Enderton, H. 86 n., 249
epistemic possibility 1, 222 ff.
equivalence relations 198 ff.
Evans, G. 164 n., 223 n., 231 n., 249

extensionality 103; crossworld, 110 ff., 230; necessity of, 105, 110

Falsehood Principle 30 ff., 32, 38, 42, 85, 106, 111, 133, 183 n.
fatalism 2, 7
Field, H. 80 n., 94 n., 226, 228, 249
Fine, K. vi, 28, 30, 35 n., 47, 49 n., 55 ff., 69 n., 75 n., 77 n., 96 n., 105, 114 n., 117, 118, 120 n., 123 n., 132 n., 172 ff., 246 ff., 249
Forbes, G. 77 n., 78 n., 151 n., 250
Four Worlds Paradox 164 ff., 185, 201

Gentzen, G. 82 n., 250
Glymour, C. 226 n., 250
Goguen, J. 170, 173, 176, 183 n., 187 n., 250
Goldblatt, R. 70 n., 94 n., 250
Grandy, R. 82 n., 250
Gupta, A. 165 n., 250

Haecceitism 148, 159, 161, 234
Hazen, A. 35 n., 66, 69 n., 89 ff., 179, 250
Hilbert, D. 94
Hintikka, J. 151 n., 248, 250
Hirsch, E. 154 n., 250
Hodges, W. 31 n., 32 n., 86 n., 250
Hughes, G. 15 n., 70 n., 250
Humberstone, L. 18 n., 21, 22 n., 45 n., 75 n., 90 n., 205 n., 231 n., 250
Hume, D. 217 ff., 220

identity 97; intrinsicness of, 139, 140 ff., 144, 188, 236; necessity of, 29, 67, 178, 179 ff., 195; personal, 130, 143, 156, 189, 193 ff.; primitive, 149 ff., 153 ff., 163; transtemporal, 52 ff., 100, 126 ff., 152 ff., 168, 190; transworld, 50, 52 ff., 100, 104, 126 ff., 149 ff., 162 ff.
indexicals 77
instrumentalism 94 ff.

Kalish, D. 72 n., 250
Kaplan, D. vi, 64, 77 n., 189 n., 197 n., 251
Kearns, J. 4 n., 251
Kripke, S. 28, 29 n., 35 n., 65, 66 ff., 69 n., 102 n., 124, 132 ff., 149, 152 ff., 179, 180, 191, 195, 204, 205 n., 222, 251

Leibniz's Law 69, 124, 125, 135, 178
Lemmon, E. 15 n., 251
Lewis, D. 53 n., 57 ff., 64 n., 69, 75, 77 n., 185, 206, 251
location essentialism 144
Locke, J. 203, 205 n.
Lombard, L. 207 ff., 251

McGinn, C. 75 n., 79 n., 134 ff., 201 n., 251
Mackie, J. 139 n., 251
Maynard Smith, J. 201 n., 233 n., 251
Mar, G. 72 n., 250
Marcus, R. 28 n., 251
materialism 26
Membership Rigidity 98, 109 ff., 123 ff., 130, 133, 134, 160, 211 n.
model for B, 15; for CTS5, 61; for S4, 16; for S5, 8 ff.; for S5BF, 55; for S5C, 54; for T, 17; for tenses, 39; homogeneous, 56
Mondadori, F. 251
Montague, R. 72 n., 250
Moravcsik, J. 2 n., 251
Morton, A. 251

natural deduction 82 ff.; natural deduction rules as meaning-giving, 87
Nozick, R. 143 n., 252

observational predicates 166 ff.
ontology 74, 81, 214

Parfit, D. 188 n., 252
Parsons, C. 114 n., 213 n., 252
Peacocke, C. vi, 85 n., 88 n., 91, 93 n., 169 n., 170 n., 219 n., 228 n., 252
Plantinga, A. 2 n., 65, 66, 75 n., 180, 252
possible worlds 8, ch. 4 *passim*, 148 ff.
Prawitz, D. 82 n., 252
Prior, A. 43 n., 75 n., 114 n., 252
propositions as sets, 10, 77 n.; individuation of, by probabilities, 226 ff.
Putnam, H. 191 ff., 198, 252

quantifiers: actualist, 34; possibilist, 34, 76, 243 ff.; substitutional, 213 n.; and tense operators, 76 ff.
quotation v ff., 50 ff.
Quine, W. V. O. 50 ff., 60, 73 n., 83, 85, 89 n., 96, 100, 119 n., 141, 163, 177, 212, 214, 216 n., 224 ff., 251

realism 74 ff., 213
Refinability 20 ff., 44
Rescher, N. 43 n., 251
rigid designator 29, 66, 183 n.
Russell, B. 72, 74, 80, 101, 108 n., 121, 251

Salmon, N. 132 n., 163 n., 165 n., 180, 196 n., 252
Schiffer, S. 29 n., 252
Schock, R. 32 n., 252
Scott, D. 15 n., 31 n., 32 n., 71 n., 120 n., 121 n., 252, 253
secondary qualities 203 ff., 218 ff.
sets ch. 5 *passim*; abstracts, 106 n., 108 n., 119 ff., 125 n; as boxes, 105, 108; cumulative hierarchy of, 102; existence of, 113; fuzzy, 174; virtual, 212 ff.; ZF theory of, 103
Sharvy, R. 124 ff., 253
Shoemaker, S. 149 n., 155 n., 235 n., 253
slime mould 160
Smorynski, C. 94 n., 253
Smullyan, A. 108 n., 253
space: logical, 77; colour, 78
Stalnaker, R. 185 n.
Stroud, B. 218, 219 n., 221, 253
substitutional semantics 88 ff.
Suppes, P. 104 n., 253

Tarski, A. 77 n., 253
Taylor, R. 2 n., 253
tenses 40 ff., 76
Thomason, R. 126
truth 83 ff.; degrees of, 170 ff.; truth functionality, 4
Twin Earth 192 ff., 198
twins, identical 128 ff.

Urquart, A. 43 n., 252

van Benthem, J. 89 n., 94 n., 253
van Dalen, D. 32 n., 253

Wertheimer, R. 167 n., 253
Wiggins, D. vi, 124 ff., 142, 147, 149 n., 221 ff., 253
Williams, S. 86 n., 88 n., 89 n.
Wittgenstein, L. 219 ff., 229, 230
Wright, C. 166, 167, 169 n., 219 n., 220, 229, 230, 253